WEATHER AND CLIMATE

WEATHER AND CLIMATE

Paul E. Lydolph
Professor of Geography
University of Wisconsin–Milwaukee

Don and Denise Temple,
Cartographers

ROWMAN & ALLANHELD
PUBLISHERS

ROWMAN & ALLANHELD

Published in the United States of America in 1985
by Rowman & Allanheld, Publishers
(a division of Littlefield, Adams & Company)
81 Adams Drive, Totowa, New Jersey 07512

Library of Congress Cataloging in Publication Data
Lydolph, Paul E.
  Weather and climate.
  Includes index.
  1. Weather. 2. Climatology. I. Title
QC981.L94 1985   551.5   84-18080
ISBN 0-86598-120-5

85 86 87 / 10 9 8 7 6 5 4 3 2 1
Printed in the United States of America

To Mary
Edward, Donald, Paul II, Thomas, Andrew
Jane, Ethel, Vicki, Joyce, Andrea
Amy, Wendy, Martha, Paul III, Tami, Mandi, Don II, Jami

CONTENTS

TABLES AND FIGURES

1 | INTRODUCTION

The temperature, moisture, and motion characteristics of the atmosphere are so all-pervading in human affairs that response to them becomes almost second nature. One is apt to forget that the climate of a place and the weather events comprising that climate overwhelmingly determine the rhythm of life. The alternation of day and night—the receipt or not of sunlight—sets the basic cycle of human existence. At breakfast each morning one turns to the radio, television or newspaper for the weather report to help plan what to wear, what precautions to take in traveling to work, and so forth. The economic fortunes of farmers completely depend on heat and moisture conditions during the growing season. The well-being of humans around the globe to a great extent depends on the availability of food and fiber and the avoidance of natural hazards such as damaging winds and floods, which are caused by the vagaries of weather and time variations of climate.

Curiosity about and dependence on the state of the atmosphere have prompted the development of broad fields of meteorology and climatology that contain within them subfields specifically conceived to serve the needs of certain facets of life and economic activity. One book of reasonable length cannot fully cover all these diverse subfields. In a single semester one can only establish a solid basis for comprehending the contemporary distribution of weather and climate over the earth and understanding the

reasons for it. Such knowledge will lay the groundwork for further studies on more-restricted topics.

This book will be limited primarily to such subfields as physical, synoptic, and regional climatology. Its two objectives are (a) to describe the climate of the earth and (b) to develop principles that will make the distribution of climate meaningful. In practice, the second objective must be accomplished first. To attach significance to regional description, one must master the atmospheric processes relating heat, moisture, and motion. Although the focus is on properties and processes at the bottom of the atmosphere at the interface between the atmosphere and the earth's surface, linkages exist that cause upper portions of the atmosphere to affect what is going on below. Therefore, to understand actions at the surface of the earth one must know something about the entire atmosphere.

To grasp the complete makeup of the climate of a place, one must know both the statistical indices of individual climatic elements and the day-to-day weather events that create these statistics. In other words, one must have a compendium of statistical means and deviations in one hand and a set of daily weather maps in the other. Places in different parts of the earth having essentially the same statistical means (for instance, of temperature and precipitation) can experience entirely different types of weather. Thus, it is not enough to deal only in statistical averages and variations. This book will consist of essentially three interwoven parts: (a) presentation of the basic principles governing atmospheric processes, (b) world distributions of statistical means and variations of individual elements and climatic types, and (c) analyses of weather complexes from which these statistics have been derived.

Although it is impossible to talk about heat and moisture without mentioning temperature and precipitation, the complete discussions of climatic elements per se have been delayed until all the atmospheric processes have been presented, since all the processes have bearings on the final, measured, climatic elements. More than heat affects temperature, and more than moisture affects precipitation. In fact, it is difficult to consider even one process at a time; so many feedback mechanisms exist in the atmosphere that almost every action causes a series of reactions. Thus, a complete consideration of temperature, probably the most commonly observed and all-pervading element of climate, is delayed until Chapter 14, after everything that affects it has been discussed.

To provide a worldwide framework within which to fit the climate of each part of the earth, Chapter 15 briefly presents a classification scheme and a distribution map of climatic types based on temperature and precipitation data. It must be realized, however, that the categories utilized in such classifications are very broad, with much variation within any given type from one continent to another. Students should remember that the types are neither immutable nor identical from region to region, and that their boundaries are only approximate. Climatic types are merely abstract mental constructs to facilitate comprehension and communication, and their boundaries are transition zones across which sweep weather events associated with atmospheric circulations whose routes of travel transgress them erratically.

The distribution of climate over the earth is presented only summarily, commensurate with the scope of the type of course for which this book is intended. For a more thorough treatment of climatic distribution, a longer version by the same author, *The Climate of the Earth,* should be used.

WEATHER AND CLIMATE

2 | CHEMICAL AND PHYSICAL STRUCTURE OF THE ATMOSPHERE

Atmospheric Composition

The earth's atmosphere is a mixture of gases in portions essentially as are shown in Table 2.1. Nitrogen makes up almost four-fifths the volume of the entire atmosphere, and oxygen more than one-fifth. All the other gases combined make up less than the remaining 1 percent. These ratios remain essentially the same to the height of more than 100 kilometers (more than 60 miles). Even so, molecular structures tend to change above 50 kilometers since exposure to direct bombardment of high-energy particles from the sun splits many molecules into their component atoms, often ionizing the air. For instance, O_2 typically splits into O_1, some of which may recombine with O_2 to form O_3, ozone. Thus, although oxygen is found at all levels, the molecule O_2 that is breathed by humans at the earth's surface is not necessarily present higher in the atmosphere.

Although nitrogen and oxygen are by far the most abundant gases, and oxygen is necessary to all life on earth, neither gas

Table 2.1 Average Composition of the Lower Atmosphere

Gas	Molecular composition	Percent by volume	Approximate molecular mass
Nitrogen	N_2	78.08	28
Oxygen	O_2	20.94	32
Argon	Ar	0.93	40
Carbon dioxide	CO_2	0.03(variable)	44
Neon	Ne	0.0018	20
Helium	He	0.0005	4
Ozone	O_3	0.00006	48
Hydrogen	H_2	0.00005	2
Krypton	Kr	Trace	
Xenon	Xe	Trace	
Methane	CH_4	Trace	
Water vapor	H_2O	Variable	18
Average			29

figures prominently in the study of climate. More important to weather processes are the heat absorption and radiation properties of gases, such as carbon dioxide and water vapor, and the consequent hydrologic cycles.

The amount of water vapor varies greatly from one place to another, according to its availability for evaporation at the surface and the holding capacity of the air above. Since all water vapor is derived from the earth's surface, it decreases rapidly with height. Unlike the other gases, water changes state within the range of temperatures commonly observed within the atmosphere; therefore, water vapor derived from the earth's surface does not accumulate in the air and mix upward thoroughly, but recondenses into liquid water and precipitates back to earth before it can rise more than a few kilometers into the atmosphere.

For these reasons, throughout much of the atmosphere there is practically no water vapor, and the average for the entire atmosphere is a small fraction of 1 percent. But near the earth's surface water vapor may account for as much as 4 percent of the atmosphere by volume in very warm, humid air. Since the water molecule is made up of two atoms of hydrogen and one atom of oxygen, its molecular mass is 18, which is considerably lighter than the molecular mass for most of the gases listed in Table 2.1. Thus, it makes up less of the atmosphere by mass than by volume, and humid air masses will be lighter than dry air masses with identical temperatures. The significance of this fact will become evident later.

Vertical Structure of the Atmosphere

Density and Pressure

Since air is a compressible mixture of gases, it is not the same density throughout. The upper air, pulled by the force of gravity toward the earth's surface, weighs down on the lower air, compresses it, and makes it more dense. Therefore, most of the mass of the atmosphere is in the bottom few kilometers. Upward from the earth's surface, the air thins very rapidly at first and then at slower and slower rates. So even though most of the atmosphere is near the earth's surface, it cannot be defined as having an outer edge. Gas molecules as high as 60,000 kilometers (37,000 miles) above the earth's surface move with the earth, attracted to it by magnetic and gravitational fields, and therefore can be considered as part of the earth's atmosphere. Of course, the air is very rare at this altitude. Half the mass of the atmosphere lies below approximately 5–6 kilometers (17,000–19,500 feet), depending on the temperature. Practically all weather phenomena with which this book will be dealing occur within the bottom 8–16 kilometers, which contain about 80 percent of the atmosphere by weight, and most of the considerations in this book will be limited to only the lower portion of that.

Since the thickness of the entire atmosphere, for practical purposes, is 100–200 kilometers, the atmosphere is only a thin shell compared to the earth, whose radius averages about 6360 kilometers. It can be seen that we are living at the bottom of a relatively shallow ocean of air, which we will consider briefly in its entirety before descending into its bottom layer, where we will be spending the rest of our time. Processes in the upper atmosphere may influence those in the lower atmosphere, although linkages are not clearly known at present.

Vertical Temperature Distribution

Evidence gathered from natural phenomena, such as the transmission of sound waves, and from direct observations by radiosondes, balloons, rockets, and satellites, have been patched together to construct a composite model of the vertical structure of the atmosphere (see Figure 2.1, Standard Atmosphere). On the average for

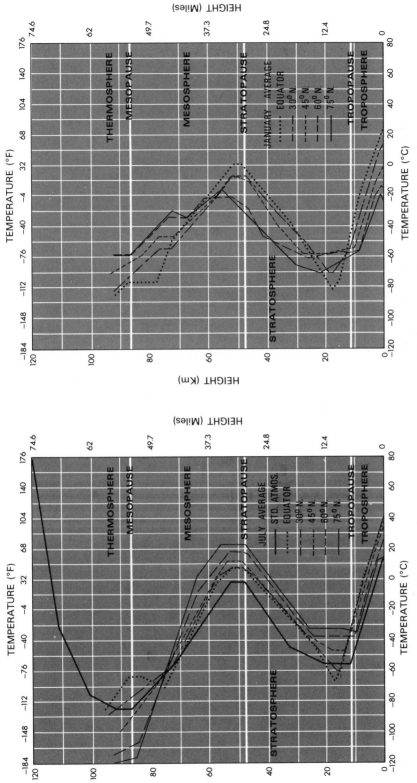

Figure 2.1. U.S. Standard Atmosphere, 1976, and mean vertical temperature distributions, July (a) and January (b) at the equator, 30°, 45°, 60° and 75° N latitude. (Data from A.E. Cole and A.J. Kantor, Air Force Reference Atmosphere, Air Force Surveys in Geophysics No. 382.)

ROBINSON PROJECTION

Figure 2.2A. Median tropopause heights (km.) during December–February.

ROBINSON PROJECTION

Figure 2.2B. Median tropopause heights (km.) during June-August. From D.F. Rex, ed., Climate of the Free Atmosphere, Vol. 4 of World Survey of Climatology (Amsterdam: Elsevier, 1969).

the entire earth, temperature at the earth's surface is around 15°C (59°F), and it decreases upward through the atmosphere at an average rate of about 6.5°C per kilometer (3.5°F per thousand feet) to an altitude of about 11 kilometers (6.8 miles), at which point the temperature is about −57°C. Above 11 kilometers it remains constant (isothermal) to about 20 kilometers and then increases. A maximum temperature of approximately −2°C is reached at around 47 kilometers, and the temperature remains at approximately −2°C up to about 52 kilometers, after which it decreases again to a minimum of about −86°C at around 86 kilometers. Above that it remains isothermal for about 5 kilometers and then increases to the outer edge of the atmosphere where it may reach temperatures of as much as 725°C or more. At this altitude the air is so rarified that lay concepts of temperature break down, as will be explained later.

Thus, the earth's atmosphere is divided into rather discrete layers with only minimal exchanges of air between them because of stability conditions, which will be explained later. The layer next to the earth's surface with a lapse (decrease) of temperature with altitude is known as the *troposphere*. This is the layer in which we live and where most weather phenomena take place. It is topped by the *tropopause* at the point where the temperature ceases to decrease and becomes isothermal. The nature and height of the tropopause has profound effects on weather phenomena in the troposphere below, and therefore a great deal will be said about it later. Above the tropopause the temperature increases in a very stable layer of air known as the *stratosphere*. The top of the stratosphere is the *stratopause,* and above that the temperature decreases through the *mesosphere* to a minimum temperature at the *mesopause,* above which the temperature increases again throughout the *thermosphere.*

Much of the vertical temperature structure of the atmosphere can be explained by considerations of absorption and transmission of heat energy originating from the sun and by temperature changes resulting from expansion and contraction of vertically moving air. (These relationships will become clear as the processes controlling them are discussed individually in succeeding chapters.) Perhaps the most-outstanding feature of the atmosphere's vertical temperature structure is the relatively warm layer in the middle, centering on about 50 kilometers height. This appears to be due to the absorption of much of the ultraviolet light from the sun by ozone (O_3), which is concentrated at and below that level. As explained earlier, ozone is one of the results of the splitting of molecules by high-energy particles in the sunlight, which causes the atmospheric molecular makeup to be somewhat different in the upper layers than it is in the troposphere at the bottom.

Other Layers

In addition to the layers named according to temperature lapse rates, certain other layers in the atmosphere are known by names that reflect molecular structures and their related phenomena such as ionization. The *ozonosphere,* for instance, extends from 15 to 50 kilometers. The *ionosphere,* where molecules have been split into their constituent atoms, with resultant negative and positive electrical charges, exists in several layers extending from about 50 kilometers to about 600 kilometers. In the far reaches of the rarefied atmosphere, between 6000 and 60,000 kilometers above the earth's surface, charged atmospheric particles, whose movement apparently is largely controlled by the magnetic field of the earth, form the Van Allen radiation belts in what is known as the *magnetosphere.* Although much remains to be learned about the connections between these upper layers of the atmosphere and the climate at the earth's

ROBINSON PROJECTION

Figure 2.3. Mean temperature, in °C, at the 200 mb. level (10500–12400 meters above sea level) in July. (From Rex, Climate of the Free Atmosphere, Elsevier, 1969.)

surface, it is obvious that the ozonosphere is very important to life at the bottom of the atmosphere because it filters out much of the harmful short wave radiation that could destroy living tissue.

Latitudinal and Seasonal Variations

As can be seen in Figure 2.1, the standard atmosphere varies significantly according to latitude and season. In the tropical atmosphere, extending about 25 to 30 degrees on either side of the equator, the troposphere is thick, the tropopause is high and consequently cold, and the temperature generally increases immediately above the tropopause rather than remaining isothermal for a while as it does at higher latitudes. The higher the latitude, the lower and warmer the tropopause and the thicker the isothermal layer in the lower stratosphere. The height of the tropopause is controlled primarily by the average temperature of the troposphere, which expands when it is warm and contracts when it is cold. Thus, outside the tropics, the tropopause is usually higher in summer than in winter.

Figure 2.2 shows these general zonal and seasonal variations of the tropopause, but it also reveals important deviations from simple latitudinal patterns, particularly in lower latitudes. The most noticeable irregularity is the great height of the tropopause during June, July, and August over southeast Asia; this is probably due to the topography of the high Tibetan Plateau, which acts as a heat source for the middle troposphere during summer. Although temperatures on the plateau surface are cool because of its elevation, they are warmer than they would be at that altitude in the free atmosphere, were not the plateau surface present to absorb sunlight.

Since the troposphere extends to greater heights in the equatorial region than in the polar areas, the normal decrease of temperature with height continues through a thicker layer in the equatorial area and eventually reaches lower temperatures at the tropopause level than is true above the poles. The tropopause is typically about $25°C$ colder above the equator than above the poles. Since the temperature eventually increases above the tropopause in the stratosphere, the lower stratosphere becomes as much as $30°C$ warmer over the poles than over the equator at the same altitude. This brings about a reversal in air flow from east to west throughout much of the stratosphere, while below, in the upper troposphere, the general circulation is from west to east.

During winter the temperature increases upward through the stratosphere faster above the equator than above the poles, so that in the middle and upper stratosphere the temperature once more becomes warmer over the equator than over the poles. But during summer the equator remains colder than the poles throughout the stratosphere. During the solstice periods throughout much of the stratosphere, there is a consistent temperature gradient from pole to pole, warm over the summer pole and cold over the winter pole, with intermediate temperatures over the equator (Figure 2.3). Thus, at these heights the equator ceases to be a climatic divide with symmetrical distributions of temperature on either side. Such a temperature distribution results in easterly winds in the summer hemisphere stratosphere and, simultaneously, westerly winds in the winter hemisphere stratosphere (see Figure 6.18). (The reason for this will be explained in Chapters 5 and 6.)

3 | GLOBAL HEAT EXCHANGE

Heat is the energy input that initiates all atmospheric processes. Its unequal distribution over the earth causes density differences in the atmosphere that bring about atmospheric motion. Heat is the energy that evaporates water from the earth's surface and causes it to mix upward into the atmosphere. Heat is intimately exchanged with other forms of energy, such as molecular motion, kinetic energy of atmospheric flow, the latent energy of the change of state of water, and the potential energy of position above the earth's surface.

For practical purposes all the heat energy that the earth's atmosphere receives originates from the sun. What happens to the sunlight as it penetrates the earth-atmosphere system depends on the characteristics of the atmosphere and the underlying surface. For the earth as a whole the entirety of climate can be explained by the amount of sunlight received and the character of the surface receiving it. For any portion of the earth-atmosphere system, however, the climate is profoundly influenced by atmospheric motion, which itself is a product of heat receipt and nature of the earth's surface. Since the heat input from the sun is the energy that initiates motion, it is important that it be considered first in considerable detail. Before that can be done, some definitions are in order.

Heat and Temperature

Heat and temperature are not the same thing, although a change in one often reflects a change in the other. Heat is a form of energy that exists in quanta that can be added or subtracted. Temperature is a measure of energy level, an expression of the mean square speed of the molecular motion of a substance. Although temperature will be dealt with in a later chapter, it is necessary to define its units of measure here, since the unit of measure for heat is based on the unit of measure for temperature.

In 1714 Gabriel Fahrenheit, of Holland, adapted the Fahrenheit temperature scale to the mercurial thermometer. He based his zero point on the lowest temperature he could obtain by mixing ice and salt. A second point on his scale was fixed by what he thought was the human temperature, which he called 96°. (This subsequently proved to be slightly inaccurate.) At any rate, on this scale the freezing point of water turned out to be 32° and the boiling point at sea level, 212°. Much of the English-speaking world adopted this scale, which has turned out to be a rather awkward one. Now the United States stands almost alone in its use, and there are plans to convert to the Celsius scale in the near future.

In 1742 Anders Celsius, a Swedish astronomer, introduced the Celsius (centigrade) scale, which has its zero degree based on the freezing point of water, and 100 degrees based on the boiling point of water at normal sea-level pressure. This has proved to be a more logical scale, particularly since the size of the degree is the same as that of the absolute (Kelvin) scale invented by Lord Kelvin (William Thompson) in 1848.

The zero point on this scale is based on the cessation of molecular motion. Since temperature is the measure of the amount of molecular motion in a substance, it is logical that with no molecular motion the temperature is zero.

Only the absolute scale has a true zero point on which to base temperature measurements that will allow additions, subtractions, multiplications, and divisions of temperatures as if they were quanta. Therefore, this scale has to be used in all scientific formulas. Since weather records around the world are all recorded in either Celsius or Fahrenheit degrees, these have to be converted first to absolute or Kelvin degrees before they can be used in formulas. Lord Kelvin chose his degree size to equal that of the Celsius degree, and therefore it is easy to convert between Celsius and Kelvin temperatures. He found that the absolute zero point was 273°C below the freezing point of water. Therefore, 0° on the Celsius scale is equal to 273° on the absolute scale, and to convert from Celsius to Kelvin all one need do is add 273 (Figure 3.1).

It is not quite so simple to convert from Fahrenheit to Celsius, since the two scales not only have different zero points, but also different degree sizes. Since it is 100°C between the boiling and freezing points of water and 180°F between the two, the Celsius degree is 1.8 times, or nine-fifths, as large as the Fahrenheit degree. To convert from Celsius to Fahrenheit, one must multiply the Celsius reading by 9/5 and then add 32. To convert from Fahrenheit to Celsius, one must first subtract 32 and then multiply by 5/9. Thus,

$$°F = 9/5°C + 32$$

and

$$°C = 5/9 \ (°F - 32).$$

Figure 3.1. Fahrenheit, Celsius, and Kelvin temperature scales.

Heat energy is expressed in a number of ways, but for the purposes of this book it will be expressed in calories.* One calorie is the amount of heat required to raise the temperature of 1 gram of water 1°C (from 14.5°C to 15.5°C). This is not the same calorie as the food calorie, which is 1000

*The international system of units (SI) recently adopted by the World Meteorological Organization (WMO) specifies the joule (J) as the official unit of energy. One calorie equals 4.186 joules or 4.186×10^7 ergs. The official unit of power (rate of expenditure of energy) is the watt (W). One watt equals one joule per second. The rate at which energy passes through or falls on a unit area is called radiant flux density. This is usually expressed in langleys per minute, where one langley equals one calorie per square centimeter (ly = cal/cm²), or in watts per square meter. One ly/min = 697.5 W/m².

times as large. To distinguish between the two, some books use the term "gram calorie" to indicate the small unit of heat that raises the temperature of 1 gram of water 1°C. Since this book always alludes to the heat unit, the simple term "calorie" will be used.

Insolation and Terrestrial Radiation

The heat received from incoming sunlight is known as *insolation*. This has been measured approximately at many different points on the earth over a great number of years and, taking into account various states of the atmosphere, it has been estimated that at the top of the atmosphere slightly less than two calories of heat are received per minute on a square centimeter of surface oriented perpendicular to the sunlight. Within the range of error of calculations, this appears to be a constant, known as the solar constant.*

The sunlight is the result of radiation of heat outward from the sun which on the average is about 150 million kilometers (93 million miles) away from the earth. Since the sun is essentially a sphere, and heat is radiated perpendicular to its surface in all directions, sun rays diverge away from the sun's surface so that the heat flux passing through any unit area decreases rapidly with distance from the sun. The rule is that the heat flux through a unit area diminishes at the rate of the distance squared that it has traveled (Figure 3.2). Thus, knowing the amount of heat received at the earth, and knowing (through astronomical measurement) the distance between the earth and the sun, one can compute the emission power of the sun. This turns out to be approximately 10^{23} kilowatts, an enormous magnitude that exceeds by many millions of times all the electrical generating capacity on earth. If

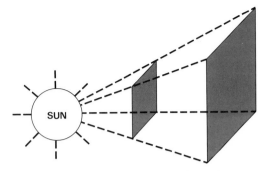

Figure 3.2. The heat radiated from the sun spreads over a larger area the farther from the sun. Thus the heat flux per unit area decreases with the square of the distance from the sun.

this amount of energy were being produced by the combustion of coal, the sun would have burned up in 5000 years, even though it is a large body with a volume approximately one million times that of the earth. The sun's energy is not being produced by the combustion of coal, however, but by the fusion of hydrogen to helium. This process will consume the sun's mass in about 14×10^{12} years, which is about 3000 times longer than the earth's estimated age at present.

By comparison to the sun and the distance between the sun and the earth, the earth is a small speck in space that intercepts only a minute portion of the total heat emission from the sun. The entire earth intercepts only about 1/2 billionth of the sun's energy output, but this still amounts to 1.8×10^{14} kilowatts, which is more than 300,000 times all the electrical generating capacity in the United States. Obviously, then, any efforts by man to add to or subtract from the heat budget of the earth by burning fossil fuels, or even by producing nuclear energy, will be puny compared to the amount of heat the earth receives from the sun every instant. Thus, man cannot hope to alter the earth's climate over broad areas by the artificial addition of heat. But man *can* hope to find ways

---

*There appears to be a little variation in solar emission, perhaps of a cyclic nature. Recent measurements from satellites have yielded an average value for the "solar constant" of 1370 W/m² (1.963 ly/min).

to divert and convert the sun's energy into more-useful forms and processes.

Knowing the emission power of the sun, one can compute the temperature of the sun's surface, since the sun radiates essentially as a black body. It has been determined in the laboratory that a black body (perfect radiator) radiates energy in proportion to the fourth power of the absolute (Kelvin) surface temperature of the body (Stefan-Boltzmann Law). Knowing the energy emission, one can solve for the temperature, which turns out to be approximately 5784°K. This, of course, is much hotter than the earth's surface, which averages about 59°F or 15°C or 288°K. Therefore, the sun emits much more heat energy than the earth does.

But the earth's temperature is a long way from absolute zero, so the earth does radiate a significant amount of heat. Most of the heat radiated to space by the earth-atmosphere system is radiated by the atmosphere at heights that are considerably colder than the earth's surface. Terrestrial (earth) radiation measured by satellites indicates that the earth-atmosphere system has an average temperature of approximately 250°K. As a result, the sun emits energy at a rate about 160,000 times that of the earth-atmosphere system.

Not only do the two bodies emit heat at very different rates, but they also emit it at very different wavelengths. All bodies not at absolute zero temperature radiate energy over a considerable spectrum of wavelengths, but this spectrum is arranged around a wavelength at which the maximum amount of energy is radiated, which has been determined to be inversely proportional to the temperature of the radiating body (Wien's Law). Therefore, the hot sun radiates its energy at much shorter wavelengths than the cool earth-atmosphere system does (Figure 3.3).

The wavelength at which the sun emits the most heat is about 0.47 micrometers (μm = 1 millionth of a meter), which is in the yellow-green portion of the visible spec-

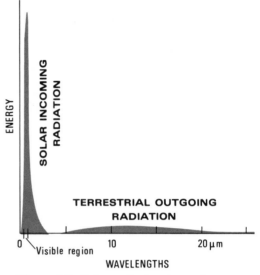

Figure 3.3. Solar radiation is much greater in magnitude and at shorter wavelengths than is earth radiation.

trum of light. Most of the sun's energy is emitted in a spectrum from 0.15μm to about 4μm. Forty-one percent of it is visible, 9 percent is ultraviolet (shorter than visible), and 50 percent is infrared (longer than visible). Terrestrial radiation, on the other hand, stretches from about 4μm to about 100μm, with maximum energy falling at about 10.1μm.

Disposition of Insolation Within the Earth-Atmosphere System

The so-called "solar constant," as mentioned before, at the outer limits of the atmosphere is approximately two calories per square centimeter per minute. But this is on a theoretical stationary surface perpendicular to the sunlight, and since the earth is a rotating sphere, sunlight hits the earth perpendicularly only at one point at any instant. Half the earth's surface is always in shadow; therefore, the average amount of heat being received on the earth's surface is considerably less than two calories per square centimeter per minute. The entire surface of the spherical earth intercepts only the same amount of sunlight as a disc the same diameter oriented perpendicular to the sunlight. Since the surface area of a sphere is four times that of a circle with the same diameter, the average amount of energy being received on the earth's surface is approximately 0.5 calories per square centimeter per minute. This, then, is the quantity of energy that one works with when considering the heat budget of the entire earth-atmosphere system. Many things happen to this energy within the atmosphere and at the earth-atmosphere interface before it is finally lost back to space again.

Scattering and absorption attenuate the sunlight by approximately half on its way down through the atmosphere. Air molecules and suspended particles scatter part of the sun's radiation in all directions. About half of this scattered light is lost back to space immediately and has no heating effect on the earth. The rest finds its way to the surface, bouncing back and forth in all directions, and provides the indirect lighting that comes through the windows of buildings and other shaded places during the day. It also lights up the daytime sky. Molecules, being very small, scatter primarily short-wave radiation and therefore cause the sky to appear blue or violet. With larger particles in the air, such as fog or smog, longer wavelengths are scattered, and the sky takes on a more whitish hue. With heavy smog, some of the light is absorbed, and the sky becomes yellowish-brown. Absorption is quite different from scattering. With absorption the air molecules take up the radiant energy and convert it to internal energy, whereby the motion of the molecules is increased or their atomic makeup is changed. This process is manifested as temperature change.

Unlike solids and liquids, which radiate and absorb heat much as a black body across a continuous spectrum of wavelengths, gases radiate and absorb heat at selected wavelengths and are transparent to others. Nitrogen (N_2), which makes up

Figure 3.4. Absorption of solar radiation. Arrow 1 represents the quantity of ultraviolet radiation at wavelengths less than 0.12μm absorbed by N_2 and O_2 in the high atmosphere above 125 km. above sea level. Arrow 2 represents quantity of ultraviolet at wavelengths between 0.12 and 0.21μm absorbed by O_2, mostly above 80 km. Arrow 3 represents quantity of ultraviolet at wavelengths between 0.21 and 0.34μm absorbed by O_3 (ozone) between 20 km. and 60 km. Arrow 4 represents quantity of near ultraviolet and visible light at wavelengths between 0.34 and 0.7μm transmitted to the earth's surface with some diminution due to scattering. Arrow 5 represents quantity of near infrared at wavelengths between 0.7 and 4.0μm transmitted to the earth's surface, absorbed slightly by O_2 and CO_2 in the upper atmosphere and more strongly by H_2O in the troposphere. (From Understanding Our Atmospheric Environment, *2nd ed., by Neiburger, M., Edinger, T.G., and Bonner, W.D., Copyright © 1982 by W.H. Freeman and Company. All rights reserved.)*

more than three-fourths of the atmosphere, absorbs only very short wavelengths, generally less than 0.12μm. Only a little of this far ultraviolet radiation is in the sunlight, and it is absorbed as soon as it enters the atmosphere by the nitrogen at high altitudes, generally above 125 kilometers. Here it dissociates molecular nitrogen, ionizes the resultant atoms, and heats the rarified air to form the so-called "thermosphere" (Figure 3.4).

Oxygen (O_2) absorbs over a somewhat

wider range of wavelengths in the ultraviolet and is also dissociated and ionized in the upper atmosphere. By the time the sunlight passes downward to about 100 kilometers above the earth's surface, all the far ultraviolet has been absorbed. Below 100 kilometers a layer extending downward to below 80 kilometers absorbs very little sunlight, because none of the radiation that is strongly absorbed reaches that low an altitude, and the density of the air is insufficient for weakly absorbed radiation to be important. The lack of absorption of much radiation centered on 80-kilometers height causes the minimum of temperature there, the mesopause (see Figure 2.1).

Below 80 kilometers a little longer-wave ultraviolet, around $0.2\mu m$, is absorbed by oxygen molecules (O_2), some of which then disintegrate into atomic oxygen (O) and mix downward to join with diatomic oxygen to produce ozone (O_3). The ozone absorbs ultraviolet radiation of wavelengths up to $0.3\mu m$, which heats up the atmosphere and produces the temperature maximum, the stratopause, at approximately 50-kilometers height. Practically all the ultraviolet that is absorbed by ozone is used up by the time it descends to 20 kilometers above the earth's surface. This is fortunate, since this short-wave radiation, if it reached the earth, would deteriorate much living tissue and perhaps make all life impossible. This is the cause of the concern about the destruction of the ozonosphere concentrated around 30 kilometers above the earth's surface by such things as supersonic transport planes that would fly in the stratosphere, and by certain chemicals in aerosol sprays that eventually mix upward to that level.

The rest of the solar radiation, between 0.3 and $4.0\mu m$, is not absorbed much by the upper atmosphere. There is some absorption of visible and infrared radiation at lower levels of the troposphere where water vapor content is significant.

The left side of Figure 3.5 diagramatically indicates the relative percentages of the different dispositions of incoming solar radiation, taking as 100 percent the 0.5 calories per square centimeter per minute that, on the average, reaches the outer limits of the atmosphere. All together the various constituents of the atmosphere just described absorb, on the average, 25 percent of the incoming solar radiation. An average of 35 percent of the incoming radiation is intercepted by clouds, which reflect about 22 percent back to space and allow 13 percent to reach the earth's surface as diffuse radiation. Another 13 percent is intercepted by atmospheric molecules that scatter approximately 5 percent of the radiation back to space and allow 8 percent to reach the earth's surface as diffuse radiation. About 27 percent of the solar radiation penetrates directly to the earth's surface, where about 24 percent is absorbed and 3 percent is reflected back to space. Hence, of the 100 percent of incoming sunlight, about 45 percent is absorbed at the earth's surface, 25 percent is absorbed in the atmosphere, and 30 percent is reflected and scattered back to space by clouds, air molecules, and the earth's surface.

The 30 percent of sunlight that is reflected and scattered immediately back to space has no effect on the earth-atmosphere system and is lost energy, except for the fact that it causes the earth to shine. This is known as the albedo (reflectivity) of the earth. Since most of the albedo is caused by reflection from cloud tops and scattering by atmospheric molecules, a planetary body without an atmosphere, such as the earth's moon, would have a much lower albedo than the earth. The moon has an average albedo of only about 7 percent. Therefore, the earth is a much brighter object in the heavens, when viewed from the moon, than the moon is, when viewed from the earth. The astronauts on their way to the moon commented on this difference in brightness of the two bodies as they looked back at the earth and exclaimed about its beauty as an object in space. On the other hand, some of the largely gaseous planets have

Figure 3.5. *Average instantaneous heat exchanges among space, the atmosphere, and the earth's surface, in percentages of incoming solar radiation (insolation).*

albedos exceeding 75 percent and appear much brighter in space than does the earth.

The 45 percent of the sun's radiation that is absorbed at the surface of the earth is used mostly to heat the surface of the earth and partially to evaporate water, which requires much heat. As the earth's temperature rises, it radiates more heat upward into the atmosphere. But since the earth's surface is much cooler than that of the sun, the heat radiated by the earth is at much longer wavelengths than that from the sun. These longer waves are unable to penetrate the atmosphere as well as the short waves from the sun, and therefore most of them are absorbed in the atmosphere. The atmosphere then warms up and radiates heat both downward and upward. The downward radiation of heat from the atmosphere adds to the heating of the earth's surface, so that at any instant the earth's surface is receiving a quantity of heat that is equal to about 140 percent of the sun's radiation. Thus the atmosphere acts as an insulating coverlet over the earth—a reservoir of heat—that is not lost immediately to space, but is absorbed and reradiated back and forth between atmosphere and earth a number of times as it slowly filters out to space. This holdover of heat in the atmosphere causes the earth's surface to be approximately 40°C warmer than it would be if there were no atmosphere and if there were an instantaneous exchange of heat between the surface and space. This entire process is known as the greenhouse effect, and it is the result of the fact that the atmosphere is not as transparent to long waves as it is to short waves.

At any instant, the earth's surface radiates upward, on the average, 111 percent as much energy as comes in at the top of the atmosphere from the sun (see right side, Figure 3.5). Most of this, 96 percent, is absorbed by the atmosphere, and only 15 percent goes directly out to space. Heat also moves upward from the earth's surface into the atmosphere through the process of evaporation of moisture (latent heat of condensation). About 18 percent is accounted for by this process. Another 11 percent moves upward as sensible heat through conduction and convection. In this process, molecules of air come into contact with molecules of earth and receive heat from them. The heated air, being lighter than colder air, moves upward in convective currents and carries the heat into the atmosphere. Thus, on the average, 140 percent as much heat as is being received from the sun is moving upward from the earth's surface, and all but 15 percent of this is being absorbed by the atmosphere. If one remembers that the atmosphere is also absorbing 25 percent of the sunlight as it moves down through the atmosphere, the total absorption by the atmosphere at any instant is $125 + 25 = 150$ percent. This heat has to be disposed of by the atmosphere, and about 95 percent, on the average, is radiated back to the earth's surface, and about 55 percent is radiated out to space. These percentages may be diminished by about 1 percent, which is transformed into the kinetic energy of motion of the atmosphere, a small fraction of which is transmitted downward to drive the ocean currents. But friction eventually reconverts this kinetic energy back to heat energy.

It is very important to remember that of the 150 units absorbed by the atmosphere at any instant, the overwhelming amount (125 percent) comes from below, and only 25 percent from above. This is perhaps the single most important fact about the earth's climate. If it were the other way around and the atmosphere were heated primarily from above, the atmosphere would be absolutely stable and would not circulate, and there would be no weather as we know it, with cloud formation, precipitation, and so forth. But as it is, the pot is constantly being boiled from below, and the atmosphere is a turbulent one.

Over the long run, there must be balances of incoming and outgoing heat at the earth's surface, at the outer limit of the atmosphere, and within the atmosphere; otherwise there would be heating or cooling trends at these levels. Figure 3.5 shows that these balances do in fact exist. At the top of the atmosphere, 100 percent radiation is received as short-wave radiation from the sun. Of this short-wave radiation, 22 percent is immediately reflected back to space by clouds, 3 percent is reflected back from the earth's surface, and 5 percent is scattered to space by atmospheric molecules. In addition to this 30 percent short-wave loss, 55 percent long waves are radiated to space from the atmosphere and 15 percent long waves from the earth's surface, for a total of 100 percent. Within the atmosphere are absorbed 25 percent of the sun's short-wave radiation, 96 percent of the earth's long-wave radiation, 18 percent latent heat of evaporation, and 11 percent sensible heat (temperature rise) due to conduction and convection between the earth's surface and the atmosphere, for a total of 150 percent. This is lost as follows: 95 percent radiated downward from the atmosphere to the earth's surface and 55 percent radiated upward to space. At the earth's surface, 45 percent of the short-wave radiation from the sun is absorbed and 95 percent of the long wave radiation from the atmosphere, for a total input of 140 percent. This is balanced by 111 percent movement upward as long-wave radiation, 18 percent as latent heat of evaporation, and 11 percent as turbulent heat through conduction and convection, for a total of 140 percent.

All the numbers in Figure 3.5 refer to radiated heat, except that being carried upward from the earth's surface by turbulent air currents, 18 percent as latent heat of evaporation and 11 percent as sensible heat. This 29 percent is referred to as the *net radiation balance* at the earth's surface, and in essence is the budget of heat energy available on the earth's surface for natural and human processes, such as the growing of plants and the weathering

of soil. In many cases, from the viewpoint of humans, it is desirable to try to increase this usable amount of heat energy. In some ways it can be altered significantly. In most cases the greatest shifts in forms of heat can probably be effected by changes in albedo. As can be seen in Figure 3.6, different types of surfaces have very different reflectivities. Fresh snow and thick clouds reflect more than 75 percent of the incident radiation, whereas wet earth may reflect as little as 5 percent. Water has the lowest reflectivity if the radiation is incident at a perpendicular angle, but at low angles water becomes quite reflective. Albedos of many types of surfaces vary significantly according to both angle and wavelength of the incident radiation. Also, cloud cover may affect the albedo of the underlying surface.

A classic example, often cited to illustrate the profound results that might be expected by changing the albedo, involves the intriguing notion of melting the Arctic ice cap. In its present state, the Arctic is a frozen sea that presents a highly reflective ice-and-snow surface to the small amount of incident sunlight reaching it during the year. Therefore, most of the small amount of heat that does reach it is immediately lost back to space through reflection. If the reflectivity of the ice-and-snow surface could be reduced by, say, covering it with coal dust or some other substance, more of the incident sunlight would be absorbed and available for melting the ice. Over a few years' time, perhaps as little as seven or eight, the Arctic ice cap, which averages only about three meters thick, might melt. Once melted, the sea would remain unfrozen because the open water surface, having a significantly lower albedo than ice and snow, would absorb enough incident sunlight to maintain its surface temperature above freezing.

This is an irreversible process, according to present technology and economic feasibility, so before one undertook such a project one should be completely sure of

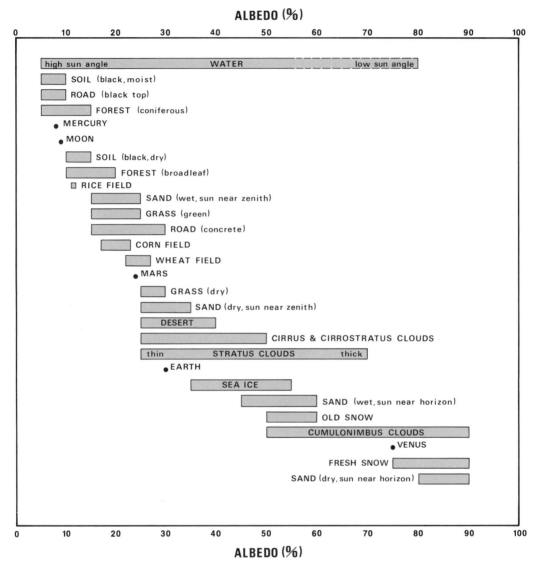

Figure 3.6. Albedos of various kinds of surfaces.

the consequences. As will be explained in Chapter 6, such a change in the Arctic might effect a circulation change in the atmosphere of the entire Northern Hemisphere, or even the whole earth, and therefore the consequences are not as simple as might at first be assumed. Not only would there be an expansion of thermal belts northward, but areas that are now humid might be turned into deserts, and so forth. Also, such a change in the Arctic would probably bring on at least partial melting of ice caps on Greenland, Antarctica, and high mountains, which would cause a catastrophic rise in sea level. Therefore, from lack of complete knowledge about the consequences of such action, no one has yet attempted such a grandiose change on the face of the earth, even though it is physically, and perhaps even economically, possible.

Of much more subtle nature, and less

risky, are heat transfers effected by irrigation, which all seem to be beneficial. Irrigation is an age-old practice used to add water to dry land. But recently it has been discovered that, in addition, irrigation normally brings about a number of beneficial and significant shifts in the heat budget at the earth's surface. By adding water to dry soil the albedo is reduced, thereby allowing for greater absorption of incoming heat energy. At the same time the temperature of the surface of the earth is lowered, because less heat goes into sensible heat and is instead transformed into latent heat of evaporation, which then is unavailable for heating the atmosphere. Since the earth is cooler, it radiates less heat, and therefore the energy budget is enhanced by a smaller loss of heat through terrestrial radiation.

If one goes to the expense of irrigating land, it is probably for growing crops on it. This will likely further reduce its reflectivity. The crops will transpire much moisture, which will increase the latent heat component of the heat transfer from earth to atmosphere. It will also transfer rather useless evaporation from a soil surface into moisture transfer through transpiration, which will effect plant growth. The crops will shade the earth, thereby further reducing the earth's surface temperature and terrestrial radiation, and will also slow down air movement so that evaporation rates are reduced. The reduction of surface temperature and motion of the air will necessitate less transpiration from each plant, so that moisture is utilized more efficiently by each plant.

Generally, plant growth is not a function of transpiration rate. Transpiration is primarily a cooling process and, if excessive, can place extreme stress on a plant because the plant is trying to evaporate water from its leaves faster than it is taking it up through its roots. The plant must then resort to a defense mechanism such as wilting, which over a prolonged period is detrimental to its growth. Of course, some transpiration of moisture through the plant is necessary to circulate nutrients from the soil upward into the plant, but this requires only a very small circulation of moisture. Thus, in most cases reduced transpiration rates are beneficial to plant growth.

In summary, then, irrigation not only adds the necessary water for plant growth, but also increases the net radiation balance at the earth's surface by cooling the surface of the earth, thereby transferring heat energy from sensible processes to latent heat processes. And much of the latent heat is transferred from simple evaporation processes to more useful transpiration processes.

The foregoing estimates of the components of the earth-atmosphere heat budget are just that and nothing more—estimates of average conditions over the entire earth. Not all the components are measurable with present technology: some have been derived from empirically determined relationships, and others have been realized only as residuals in equations. The latter might encompass considerable error if errors were inherent in measurements of the other variables, since the residuals would absorb the sum of the errors. Thus, magnitudes will be found to vary from one textbook to another, but the nature of the variables themselves are the same, and the quantitative estimates are better than mere guesses.

4 | SPATIAL AND TEMPORAL HEAT BUDGET VARIATION

The heat budget components vary greatly from place to place and from time to time at any one place. Insolation varies by latitude because the earth is a sphere. It varies seasonally at any one place because the earth is rotating on an axis, revolving about the sun, and the axis of rotation is inclined with respect to the orbit around the sun. By the same token, the insolation varies with time of day because the sun comes up in the east, crosses the zenith at noon, and sets in the west, thereby causing the angle of incidence at any point on the earth to vary constantly during the day.

Undulations in the earth's surface, of course, vary the amounts of heat received because of different angles of slope. A north slope obviously receives less heat than a south slope in the middle and high latitudes of the Northern Hemisphere. In the morning an east slope will receive more heat, and in the afternoon a west slope.

In addition to all these geometrical relationships, the irregularities of atmospheric conditions, particularly clouds and humidity, as well as the characteristics of the receiving surface, complicate the pattern immeasurably. Since the earth's surface is approximately three-fourths water, the widely varying albedo of water with angle of incident sunlight is an important variable.

Sun-Earth Relations

First, consider the regularly varying spatial characteristics brought about by the spher-ical shape of the earth and its rotation and revolution about the sun. The earth revolves around the sun approximately once per year in a slightly elliptical orbit with the sun at one focus of the ellipse. The distance between the earth and the sun averages about 150 million kilometers (93 million miles). This distance varies by about 1.7 percent. The greatest distance, *aphelion*, occurs on 5 July, and the shortest distance, *perihelion*, occurs on 4 January. This variation in distance causes a plus or minus 3.4 percent variation in heat receipt by surfaces perpendicular to the sun's rays at the outer limit of the atmosphere. At the surface of the earth, however, local factors generally mask this variation.

Angle of Incidence

Much more important than distance in determining the amount of heat received is the curvature of the earth's surface, which varies the angle at which the sunlight falls on the earth. Since the earth is so far away from the sun and such a small speck in space compared to the sun, the small amount of sun's energy that is incident upon the earth's surface has been radiated from only a very small part of the sun's surface, which for practical purposes can be considered as a flat surface, and the rays then can be considered to be parallel to each other. Therefore, the angle at which the sun's rays strike the earth's surface depends essentially on the latitude of the earth and is a complement to the angle of latitude. That is, if the sun is directly overhead at the equator, which is 0° latitude, the angle of incidence will be 90°. At 30° latitude the angle of incidence will be 60°, at 45° latitude, 45°, and so forth. At the pole, 90° latitude, the angle of incidence will be zero.

A direct ray yields more heat per unit area than an oblique ray does, for three reasons. First, the vertical ray will spread over a smaller area when it hits the earth's surface than a slanting ray will. Second, the slanting ray will pass through a greater thickness of atmosphere than the vertical ray and therefore become more attenuated on its way through the atmosphere. When the sun is only 5° above the horizon, its rays must travel through almost eleven times as much atmosphere as when the rays are perpendicular. The radiation intensity on a horizontal surface, then, is little more than 1 percent what it is when the rays are perpendicular (Table 4.1). And third, the albedo over many surfaces is greater for slanting rays than for direct rays.

In addition to the facts that the earth is spherical and is revolving around the sun, it is also spinning on an axis once a day, and this axis of rotation is tilted at an angle approximately 23.5° away from the perpendicular to the orbit of revolution around the sun. This tilt of the axis remains essentially parallel to itself as the earth revolves about the sun, so that at different points in the orbit of revolution different portions of the earth's surface are exposed to sunlight (Figure 4.1). This combination of circumstances produces the seasons.

On approximately 21 June the north pole is tilted most directly toward the sun, and the sun's last tangent rays pass across the pole to 66.5°N latitude on the other side (Figure 4.2). At this time the direct rays of the sun are overhead at 23.5°N latitude. On 21 December the situation is reversed: the south pole is tilted toward the sun, the north pole away from it, and the last tangent rays of the sun in the Northern Hemisphere

Table 4.1 Atmospheric Thickness and Radiation Intensity on a Horizontal Surface for Varying Sun Altitude above the Horizon, Assuming a Transmission Coefficient of 78 Percent

Sun's altitude	Number of atmospheres rays must pass through	Radiation intensity (in percent)
90°	1.00	78
80°	1.02	76
70°	1.06	72
60°	1.15	65
50°	1.31	55
40°	1.56	44
30°	2.00	31
20°	2.92	17
10°	5.70	5
5°	10.80	1
0°	45.00	0

reach only to 66.5° latitude. The direct rays of the sun are now overhead at 23.5°S latitude. These extreme positions of the direct and tangent rays of the sun set the positions of the Tropic of Cancer at 23.5°N latitude, the Tropic of Capricorn at 23.5°S latitude, the Arctic Circle at 66.5°N latitude, and the Antarctic Circle at 66.5°S latitude.

Daylight Period

Not only does the angle vary at which the sun's rays are incident upon the earth's surface, but also the length of daylight varies with latitude because of the tilt of the axis. As can be seen in Figure 4.2, the circle of illumination, which divides the earth into a light half and a dark half, includes the entire North Polar area southward to the Arctic Circle within the light half on 21 June (summer solstice) and within the dark half on 21 December (winter solstice). This means that on 21 June the pole and all the area around it south-

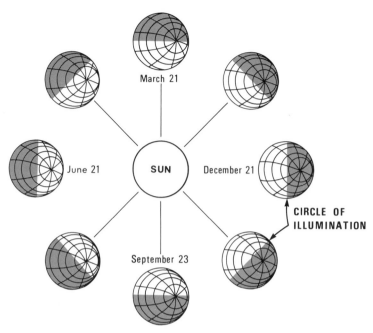

Figure 4.1. Changes in earth's position relative to the sun during the course of a year.

ward to the Arctic Circle receives sunlight all 24 hours of the day, while on 21 December this area receives no direct sunlight at all. On 21 March and 23 September the circle of illumination passes through the poles, and therefore all points on the earth's surface have 12 hours of daylight and 12 hours of darkness on these dates, which are called the vernal (spring) and autumnal (fall) equinoxes.

The length of daylight period at any place at any date can be judged by the portion of the latitudinal parallel that falls within the lighted half of the earth relative to the portion that falls within the darkened half (Table 4.2). At the equator there are 12 hours of daylight and 12 hours of darkness every day of the year. This varies more widely the farther one goes toward the poles, and at the poles themselves there is a six-months' daylight period followed by a six-months' darkness period. At intermediate positions between the equator and poles, the length of daylight period increases rapidly during spring, then levels off as the summer solstice is approached, falls rapidly during autumn, and tapers off again as winter is approached (see Table 4.2). For instance, at 40° latitude, the daylight period lengthens from 9 hours 37 minutes on 15 January to 10 hours 42 minutes on 15 February, a change of 1 hour 5 minutes during the month. It lengthens from 11 hours 53 minutes on 15 March to 13 hours 14 minutes on 15 April, a change of 1 hour 21 minutes during that month. From 15 May to 15 June it lengthens from 14 hours 22 minutes to 15 hours, a change of only 38 minutes.

Latitudinal Effects on Insolation Receipts

Latitude, then, exercises two variables on insolation receipt, angle of incidence and length of daylight period. These sometimes vary with latitude in the same direction and sometimes in opposite directions to produce a composite latitudinal heat distribution as shown in Figure 4.3. In the

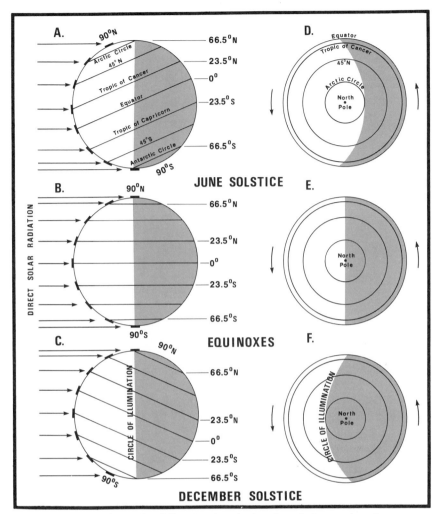

Figure 4.2. The circle of illumination during the solstices and equinoxes. (Modified from John B. Leighly.)

Northern Hemisphere on 21 June (s) the sun at noon is directly overhead at 23.5°N latitude. The elevation above the horizon (H) decreases in both directions from 90° at the Tropic of Cancer to 66.5° at the equator and 23.5° at the pole. The length of day (D) varies with latitude from 12 hours at the equator, slowly at first and then more rapidly, until it reaches 24 hours at 66.5°, at which it remains all the way to the pole. These two curves, marked H_s and D_s in the diagram, add up graphically to I_s, the amount of insolation received at the outer limit of the atmosphere. Since from the equator to the Tropic of Cancer, both the angle of incidence and the daylight period increase initially, the amount of insolation received increases with latitude. But shortly poleward of the Tropic of Cancer, the insolation begins to decrease slightly because the angle of incidence is decreasing faster than the daylight period is increasing. Poleward of about 50° latitude, however, the daylight period increases more rapidly than the angle of incidence decreases, and the insolation receipt increases again all

Table 4.2   Length of Daylight Period by Latitude on the 15th of Each Month, in Hours and Minutes (not including light refracted by the atmosphere)

	0°	10°	20°	30°	40°	50°	60°	70°	80°	90°
Jan.	12:00	11:35	11:02	10:24	9:37	8:30	6:38	0:00	0:00	0:00
Feb.	12:00	11:49	11:21	11:10	10:42	10:07	9:11	7:20	0:00	0:00
Mar.	12:00	12:04	12:00	11:57	11:53	11:48	11:41	11:28	10:52	0:00
Apr.	12:00	12:21	12:36	12:53	13:14	13:44	14:31	16:06	24:00	24:00
May	12:00	12:34	13:04	13:38	14:22	15:22	17:04	22:13	24:00	24:00
June	12:00	12:42	13:20	14:04	15:00	16:21	18:49	24:00	24:00	24:00
July	12:00	12:40	13:16	13:56	14:49	15:38	17:31	24:00	24:00	24:00
Aug.	12:00	12:28	12:50	13:16	13:48	14:33	15:46	18:26	24:00	24:00
Sept.	12:00	12:12	12:17	12:23	12:31	12:42	13:00	13:34	15:16	24:00
Oct.	12:00	11:55	11:42	11:28	11:10	10:47	10:11	9:03	5:10	0:00
Nov.	12:00	11:40	11:12	10:40	10:01	9:06	7:37	3:06	0:00	0:00
Dec.	12:00	11:32	10:56	10:14	9:20	8:05	5:54	0:00	0:00	0:00

the way to the pole. The pole at this time of the year at the top of the atmosphere receives more heat energy over a 24-hour period than any other latitude.

On 21 December (w), on the other hand, the sun is directly over the Tropic of Capricorn, 23.5°S latitude, and the equator again experiences a sun that at noon stands only 66.5° above the horizon. The length of the daylight period at the equator is 12 hours, as is true throughout the year. Both the angle of incidence, H_w, and the length of daylight period, D_w, decrease rapidly from the equator to 66.5°N latitude, where they both become zero and remain at that level all the way to the pole. Thus, in the Northern Hemisphere the insolation receipt, I_w, decreases steadily from a maximum over the equator to zero at the Arctic Circle and beyond. Insolation over the equator at this time of year is slightly greater than it was on 21 June, because the earth is now closer to the sun.

It is evident that in the middle latitudes, the latitudinal changes of daylight period and angle of incidence are in the same direction during winter but in opposite directions during summer. Thus, latitudinal differences in insolation are much greater

during winter than during summer. Hence, latitudinal changes in temperature are much greater in the middle latitudes during winter, which leads to a more vigorous atmospheric circulation during winter.

At the equinoxes (E), every latitude receives 12 hours of daylight and 12 hours of darkness. The sun is directly overhead at the equator, and the angle of incidence, H_E, decreases consistently from 90° at the equator to zero at the pole. Since the length of daylight period, D_E, does not vary at this time of year, the amount of insolation received, I_E, varies directly with the angle of incidence of the sunlight on the earth.

The longest and shortest days of the year are known as the summer and winter solstices, respectively. In the Northern Hemisphere these fall approximately on 21 June and 21 December. In the Southern Hemisphere they are just the reverse. Since the earth does not revolve around the sun in exactly 365 days, every fourth year is a leap year (has an extra day). Thus, the dates of the solstices and equinoxes vary one or two days from year to year. Also, the earth's axis itself rotates slowly during periods of about 26,000 years. Therefore, the seasons precess slowly through time. In 13,000 years,

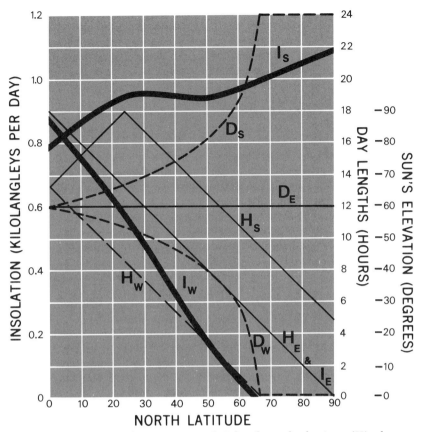

Figure 4.3. Angle of sun's rays or height above the horizon (H), day length (D), and insolation receipt (I) at the top of the atmosphere, by latitude, during the summer (s) and winter (w) solstices and the equinoxes (E).

the seasons will be just the reverse of what they are now: the Northern Hemisphere winter solstice will fall in June and the summer solstice in December.

It happens that at present the summer solstice of the Northern Hemisphere nearly coincides with the position of the earth at aphelion (farthest point from the sun), and the winter solstice nearly coincides with the timing of the earth in its position at perihelion (closest to the sun). The angle of tilt of the pole toward the sun and the nearest and farthest distances of the earth in its orbit from the sun are not one and the same thing, and there is no reason for them to coincide in time. As the seasons

precess through the years, this coincidence will become less marked.

The Southern Hemisphere, of course, is closest to the sun during its summer season and farthest from the sun during its winter season. Knowing this one fact, one might conclude that the Southern Hemisphere would have greater seasonal extremes than the Northern Hemisphere does. This, in fact, is true in the case of insolation receipt. As far as resultant surface temperatures are concerned, however, the situation is the reverse. The greater preponderance of water in the Southern Hemisphere tempers seasonal extremes there.

Figure 4.4 shows the distribution of insolation at the outer limit of the atmosphere by month and latitude. It can be seen that the greatest values on the entire earth are realized over the poles during their short summer seasons, and the South Pole receives slightly more insolation during its summer than the North Pole does during its summer, because of the ellipticity of the orbit. The equator varies the least through the year and receives its maximum insolation during the equinoxes. Over the course of the year the equator receives considerably more heat than any other latitude.

If the effects of a clear atmosphere are added, assuming 70 percent transmission through the normal thickness of the atmosphere, the distribution of insolation changes significantly (Figure 4.5). Since the slanting rays in the high latitudes must pass through more atmosphere, the insolation is depleted more at high latitudes than at low latitudes. Heat receipt is much less everywhere on the earth's surface with the atmosphere than without it, but it reduces from more than 44 megajoules/square meter (1000 langleys) per day to only about 18 MJ/m² (450 langleys) per day at the pole, while it declines only from about 40 MJ/m² (925 langleys) per day to about 25 MJ/m² (575 langleys) per day at 30° latitude during the summer solstice. The result is that the maximum heat receipt at the earth's surface during the solstice periods occurs in the middle latitudes centered on about 30°–35° latitude. Maximum heating at the earth's surface thus shifts during the year over a span of latitude averaging about 65°, more than the 47°-latitude shift of the direct rays of the sun.

Atmospheric Effects

This is still a very hypothetical picture, since it does not take into account the state of the atmosphere, primarily clouds and moisture content. The assumed transmissivity of 70 percent is frequently exceeded.

Also, sky (scattered) radiation is added to direct radiation, so that total daily global radiation in excess of 30 MJ/m² (720 langleys) is not rare even at low altitudes, and mountain stations may experience even higher values. On cloudy days, of course, transmissivity is reduced. Figure 4.6 depicts the pattern of insolation receipt at the earth's surface as actually measured. Here the pattern is far from a purely zonal (latitudinal) one. The greatest amounts are received in such desert areas as the Sahara of northern Africa, the Kalahari of southern Africa, central Australia, and southwestern United States and northwestern Mexico. Here the skies are generally clear and the air relatively free of moisture. The equator, on the other hand, which receives the most heat during the year at the outer limits of the atmosphere, has considerable heat attenuation due to great amounts of cloudiness and high humidity content. (The reasons for such distributions of clouds and humidity will become clear later when atmospheric circulation and resultant cloud conditions are discussed.) Seasonal patterns of insolation are similar to the annual pattern, with some latitudinal shifts corresponding to shifts in the angle of incidence of the sun and the length of daylight period.

Heat Exchange at the Earth's Surface

Radiation Balance

The same dry atmospheric conditions that allow maximum insolation over dry land areas also allow maximum terrestrial radiation to escape to space, and therefore these areas utilize the heat they receive very inefficiently and show up as minimum areas of radiation balance at the earth's surface (Figure 4.7). The world pattern of radiation balance, then, is much different from that of radiation receipt. Whereas land areas, with their lower atmospheric mois-

Figure 4.5. Insolation receipt on a horizontal surface at sea level, assuming 70 percent transmission through a normal atmosphere, by month and latitude, in MJ/m² day. (From Understanding Our Atmospheric Environment, 2nd ed., by Neiburger, M., Edinger, T.G., and Bonner, W.D. Copyright © 1982 by W.H. Freeman and Company. All rights reserved.)

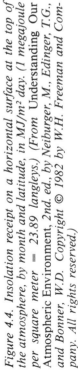

Figure 4.4. Insolation receipt on a horizontal surface at the top of the atmosphere, by month and latitude, in MJ/m² day. (1 megajoule per square meter = 23.89 langleys.) (From Understanding Our Atmospheric Environment, 2nd ed., by Neiburger, M., Edinger, T.G., and Bonner, W.D. Copyright © 1982 by W.H. Freeman and Company. All rights reserved.)

Figure 4.6. Distribution of actual average annual insolation at the surface of the earth, in kilolangleys. (From M.I. Budyko, ed., Atlas teplovogo balansa zemnogo shara [Atlas of Heat Balance of the Earth] [Moscow, 1963].)

ROBINSON PROJECTION

MOUNTAINS

0   1500   3000 KMS

Figure 4.7. *Average annual radiation balance at the surface of the earth, in kilolangleys. (After* Budyko, **Atlas of Heat Balance of the Earth.**)

MOUNTAINS

ROBINSON PROJECTION

0   1500   3000 KMS

ture contents, generally receive more insolation than adjacent water areas and the highest values of all are in the large deserts of the world, the radiation balance pattern generally shows higher values over water than over adjacent land, and the lowest values of all occur over the great subtropical deserts, such as the Sahara and Australia, or in the earth's high-mountain regions, where the air is thin and low in moisture. The deserts and high mountains of the earth have often been called the "windows of the world," because both areas are associated with atmosphere that allows ready passage of both incoming and outgoing radiation.

In general, oceans absorb much more of the incoming solar radiation than land surfaces, and radiate much less back to space because of their consistently moderate temperatures. The oceans thus show a high radiation balance and serve as a huge reservoir for heat storage on the earth.

It is interesting to note that over the entire course of the year, the radiation balance at the earth's surface almost everywhere is positive. The pattern of radiation balance, of course, changes drastically with seasons. From May through August all the earth's surface, except for the Southern Hemisphere south of about 42°S latitude and perhaps parts of the Arctic, shows a positive radiation balance. By September a negative balance begins to show up on the northern fringes of the Canadian Archipelago and the Arctic islands off Siberia. This spreads rapidly southward in succeeding months and reaches a latitude of about 40°N (37°N over Japan) at its maximum extent during the Northern Hemisphere winter solstice in December. The area of negative radiation balance begins to recede northward again in January, and by April it has practically disappeared off the Northern Hemisphere continents except for the northern islands in the Canadian Archipelago and Spitsbergen, Franz Joseph Land, and Severnaya Zemlya in the Arctic north of Eurasia.

Latent Heat Exchange

As might be expected, much of the radiation balance over the oceans is used as latent heat to evaporate water, particularly in subtropical latitudes where the atmosphere is usually clear of clouds and relatively dry (Figure 4.8). Values are even higher in certain ocean areas adjacent to middle latitude continents, such as off the east coast of the United States and portions of the western North Pacific next to Japan. In these middle-latitude areas the excessive evaporation takes place primarily during winter. (The reasons for this will be discussed in Chapter 7.) The land masses, where moisture generally is not as readily available to be evaporated, show much lower values, although some of the equatorial land areas, such as the Amazon Basin, the Congo, and the Indonesian Islands, still show relatively high values.

Sensible Heat Exchange

The sensible heat exchange map is almost the exact complement of the evaporation map (Figure 4.9). Heat is moved upward through conduction and convection (as an air temperature increase) most rapidly where the earth's surface is hottest—the earth's subtropical deserts, such as the Sahara and Kalahari in Africa and the large Australian desert. The oceans, with their relatively cool summer temperatures, engage in very little sensible heat exchange with the atmosphere above them. A few oceanic areas, such as the North Atlantic and western North Pacific, where large, warm ocean currents carry anomalously warm waters into middle and higher latitudes, show high values of sensible heat exchange, particularly in winter. The Barents Sea north of Scandinavia is outstanding in this respect.

Figure 4.8. *Average annual amount of heat used to evaporate water (latent heat), in kilolangleys.*
(After Budyko, Atlas of Heat Balance of the Earth.)

MOUNTAINS

ROBINSON PROJECTION

0   1500   3000 KMS

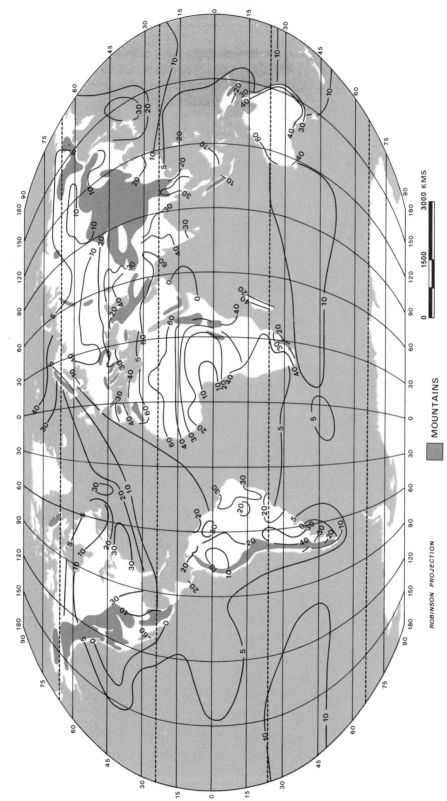

Figure 4.9. Average annual sensible heat exchange between the earth's surface and the atmosphere, in kilolangleys. (After Budyko, Atlas of Heat Balance of the Earth.)

MOUNTAINS

ROBINSON PROJECTION

0   1500   3000 KMS

On the other hand, some ocean areas with cool surface currents experience a sensible heat exchange that is directed downward from the atmosphere to the water.

Heat Balance of the Earth-Atmosphere System

Over the course of a year, the net radiation balance at the earth's surface is positive over most of the earth. Therefore, the net radiation balance of the atmosphere must be negative for the radiation balance of the earth-atmosphere system to be zero so there is no net heating or cooling over a long period of time. Most of the heat lost to space is radiated from the atmosphere, particularly from cloud tops. The atmosphere thus has a radiation balance of approximately -80 kilolangleys per year at every latitude. It varies slightly by latitude, but not very much. The atmosphere, of course, makes up for this heat deficit by receipt from the surface of the earth.

Horizontal Transfers of Heat

The earth-atmosphere system has a net gain of heat from approximately 32°N latitude to 28°S latitude, and a net loss poleward from those latitudes. Figure 4.10 shows these relationships for the Northern Hemisphere. The latitudinal scale is a cosine one in order to represent equivalent areas of the earth's surface properly. Since the parallels of latitude become shorter from equator to pole, there is less area between equal latitudinal intervals near the poles than near the equator. Half the earth's surface falls between 30°S latitude and 30°N latitude. In Figure 4.10 it can be seen that the surplus in the low latitudes balances the deficit in the middle and high latitudes.

For the low latitudes to get rid of surplus heat and for the high latitudes to radiate more heat to space than they receive from the sun, there must be a horizontal transfer of heat from low latitudes to high latitudes.

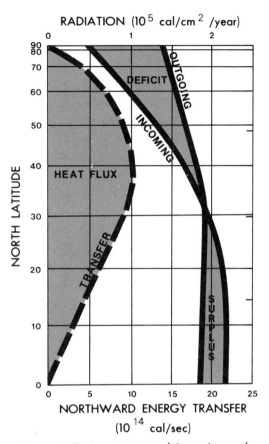

Figure 4.10. Average annual incoming and outgoing radiation for the earth-atmosphere system and the poleward transfer of heat.

The dashed line in Figure 4.10 indicates roughly the magnitude of the heat flux by latitude. It is obvious that the greatest heat flux is in the middle latitudes around 30°–40°, for all the heat that moves from low latitudes to high latitudes must pass through the middle latitudes. This necessitates a maximum of air movement in the middle latitudes to effect the heat transfer. In Chapters 6 and 12 it will become evident that the middle latitudes, on the average, have the highest wind speeds and are the stormiest and in general, climatically most vigorous of the earth's zones.

The transfer of heat from low latitudes to high latitudes is accomplished by three

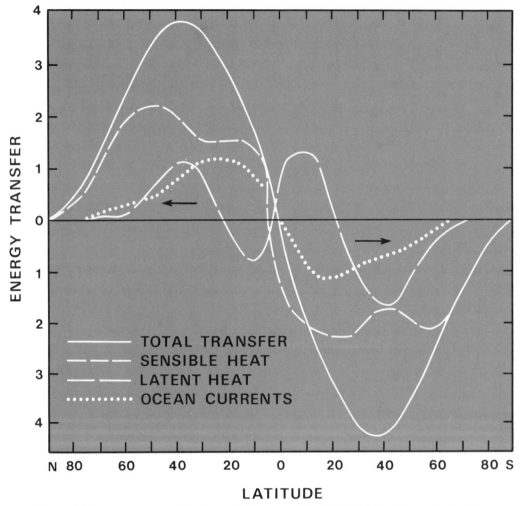

Figure 4.11. Average annual horizontal heat transfers, in 10^{19} kilocalories. (After W.D. Sellers, Physical Climatology *[Chicago: University of Chicago Press, 1965]. © 1965 by the University of Chicago. All rights reserved.)*

means (see Figure 4.11). The largest component is sensible heat transfer in the atmosphere; that is, the movement of warmer air into higher latitudes. In each hemisphere there is a fairly simple poleward movement, with maximum flux taking place in the middle latitudes. Ocean currents tend to show the same pattern of heat transfer, at lesser magnitudes. The movement of latent heat via water vapor, however, shows a different pattern. Since most of the evap-

oration on earth takes place over the subtropical oceans, latent heat moves both equatorward and poleward from the subtropical areas on either side of the equator. Since the magnitude of latent heat movement is considerably smaller than that of sensible heat movement in the atmosphere and oceans combined, this more complicated pattern of latent heat transfer does not perceptibly affect the overall pattern of heat transfer.

5 | ATMOSPHERIC PRESSURE AND MOTION

Since different amounts of heat energy are received at different points on the earth's surface, and since these inputs change from time to time at any one place, differences in density are set up in the earth's atmosphere that cause atmospheric motion. Heat energy is the driving force that initiates the motion, but once in motion, the atmosphere is acted upon by such forces as that due to the rotation of the earth, and a circulation evolves that tends to maintain itself and take on a character of its own, which in turn influences air pressure and temperature. Thus, on any portion of the earth's surface, atmospheric motion and the heat budget must be considered as major controls on climatic elements. In the middle latitudes, particularly, the surface air temperature on any given day is probably due more to movements of air over the region than to the radiation balance at that point on the earth's surface. The same is true of atmospheric humidity, cloud formation, precipitation, and so forth. Therefore, a knowledge of the movements in the atmosphere and their consequences is all-important. Since this motion is a result of atmospheric pressure differences, it is necessary to define atmospheric pressure.

Air Pressure

The mixture of gases that make up the atmosphere has a certain mass and, al-though not very dense, adds up to considerable weight when one considers the entire thickness of the atmosphere attracted by the force of gravity to the earth's surface. On the average at sea level, the atmosphere weighs approximately one kilogram per square centimeter (14.7 pounds per square inch). This is quite a great weight if one considers how many square centimeters are on, say, the surface of the human body or the roof of a building. For instance, the ceiling of a room 3 × 4 meters square would contain 120,000 square centimeters, which would be under an atmospheric weight of 120 metric tons. Of course, in a gas, pressure is exerted equally in all directions, and as long as there is as much air pressure inside a room as outside there is no net force on the structure. But if the air were suddenly evacuated from the room, it is easy to see that the structure would collapse under the great weight of the atmosphere.

Measurement and Expression

The most accurate instrument for measuring air pressure is the mercurial barometer, which consists of an evacuated glass tube about one meter long, mounted vertically, with a surface of mercury at the base. The weight of the air pressing down on the mercury surface balances a column of mercury in the tube equal in weight to the weight of the atmosphere. Thus, air pressure is often given in terms of the height of the column of mercury balanced by the air. Normal sea-level pressure in these terms is 76 centimeters or 29.92 inches.

Another kind of instrument is the aneroid barometer, which consists of an evacuated metal wafer that expands and contracts with air pressure and moves a pointer across a scale. Since this is entirely mechanical, contains no liquids, and can be moved easily and held in any position, it is much more convenient to use. It is not as accurate as the mercurial barometer, however. Since, for most purposes, not the absolute pressure but the relative pressure from one place to another or the change at a given point over time is important, an instrument called a barograph was invented to show a continuous trace of air pressure over time. This uses an aneroid barometer to move a pen arm on a graph paper wrapped around a slowly rotating cylinder, thereby tracing a continual record of the changes of air pressure.

Neither pounds per square inch nor inches or centimeters of mercury are very convenient for plugging into scientific equations, and therefore most of the world has gone to the expression of air pressure in millibars (mb). Since one millibar equals 1000 dynes per square centimeter, and one dyne equals one gram-centimeter per second per second, the millibar is based on the c.g.s. system and is convenient to use in scientific equations. Average sea-level pressure in millibars is 1013.2. Almost all atmospheric pressures recorded in the United States since 1940 have been expressed in millibars, as is true in most of the rest of the world.* Nevertheless, airports still report altimeter settings in inches of mercury, so it is important to be able to convert from one to the other. One inch

of mercury equals approximately 34 millibars.

Variation

Since the atmosphere is usually involved in some motion, and different amounts of heat energy are being expended in different ways, atmospheric pressure undergoes constant variation at any given point on the earth's surface and varies from place to place at any given time. In a horizontal direction, the variations are not great. At sea level the variation is usually no more than 5 percent on either side of the normal, which is probably not enough for the human body to detect consciously. Air pressure varies much more rapidly with height, however; as was pointed out before, the air pressure decreases by half about every 5000–6000 meters. Therefore, roughly half the mass of the atmosphere lies below 5500 meters (18,000 feet). Anyone who has climbed a high mountain or flown in an unpressurized airplane knows that changes in air pressure of this magnitude can cause extreme reactions in the human body, ranging from accelerated breathing and general fatigue in milder cases, to severe headaches, nosebleeds, nausea, dizziness, and fainting, in more severe cases.

Although the horizontal variations in air pressure might not be of physiological significance, they are all-important to atmospheric motion. And since atmospheric motion is such an important control on all other weather phenomena, the distribution of air pressure over the earth's surface at any given time is probably the single most important tool for predicting a whole range of weather elements. Because the motion itself, once initiated by heat differences, becomes a control on the air pressure pattern, there is a feedback mechanism be-

---

*The recently adopted SI system specifies the kilopascal (kPa) as the official unit of air pressure. One kPa equals 10 mb. One pascal (Pa) equals one newton (N) per square meter, where one newton is the force that, acting on a mass of one kilogram, produces an acceleration of one meter per second per second.

tween the two that requires air pressure and motion to be considered jointly. First to be considered are broad-scale patterns of pressure and motion over the earth, and then smaller circulation systems. So we may grasp the workings of the general circulation of the atmosphere, a simple case will be considered first, and then complications will be added one at a time until reality is approached.

Atmospheric Circulation

Wind-Force Relations

Let us begin with a homogeneous, stationary earth receiving the sun's energy on its spherical surface. Under these conditions, the equatorial region would receive more heat than the polar areas, and an air pressure gradient (decrease) would be set up at the earth's surface directed from the cold, dense air over the pole to the warm, light air over the equator. The resultant circulation theoretically would be a single vertical cell between equator and pole (Figure 5.1). The warm, light air at the equator would rise, and the cold, dense air at the pole would sink. This would induce a pole-to-equator movement of air at the surface and an equator-to-pole movement of air aloft.

Coriolis Force

The earth is not just sitting in space, however, but is rotating on an axis, and therefore the north-south-moving air streams are flowing above a rotating plane of reference from which the motion is being judged. An observer rotating with the earth's surface would perceive the air streams to curve with respect to this frame of reference even though they might appear to an observer in space to be straight lines (Figure 5.2). Since the earth rotates from west to east, it is rotating counterclockwise as viewed from above the North Pole. Therefore, in

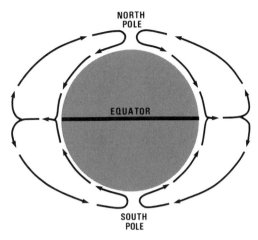

Figure 5.1. Single-cell hemispheric circulations on a homogeneous, stationary earth.

the Northern Hemisphere this apparent deflection of an object moving across the face of the earth appears to be clockwise, or to the right of the motion, viewed downstream with the motion. As viewed from the South Pole, the earth appears to be rotating clockwise, so in the Southern Hemisphere the deflection appears counterclockwise, or to the left of the motion, as viewed downstream. This apparent "force" is known as the Coriolis Force after G.G. de Coriolis (1792–1843), a French mathematician who first treated the phenomenon quantitatively.

It can be shown by such devices as Foucault pendulums (heavy balls on long, thin wires) in museums that because of the earth's rotation about its axis, any point on the surface of the earth acts as a center of rotation for its own horizon about an axis perpendicular to the earth's surface at that point. The speed of rotation of the horizon depends upon the latitude, increasing from zero at the equator to a maximum at the poles. Thus, the Coriolis Force also increases from zero at the equator to a maximum at the poles. In addition, since the Coriolis Force is a relative motion between a moving object and the rotating surface of the earth, the Coriolis Force also increases with the speed of the moving object. No matter which direction the motion takes, the Coriolis Force is directed perpendicular to the motion, to the right in the Northern Hemisphere and to the left in the Southern Hemisphere.

Thus there are two basic forces acting upon the air, the pressure gradient force (pressure decrease from one place to another) and the Coriolis Force. As long as these two forces are out of balance with each other, the air will change its speed and/or direction, and by so doing quickly bring into balance the two forces so that the air will move with a constant velocity (speed and direction).

Figure 5.3 illustrates in slow motion what is essentially an instantaneous process of adjustment. Assume that over a small portion of the earth's surface in the Northern Hemisphere the upper air contours (lines of equal altitude on a surface of constant pressure) are straight and equally spaced, as shown in Figure 5.3. The pressure gradient force, P, is directed from high pressure to low pressure. The air begins to move down this pressure gradient, but as soon as motion begins the Coriolis Force, C, deflects the air to the right. As long as the air is moving partially in the direction of the pressure gradient, it will accelerate, and the faster it goes, the stronger the Coriolis

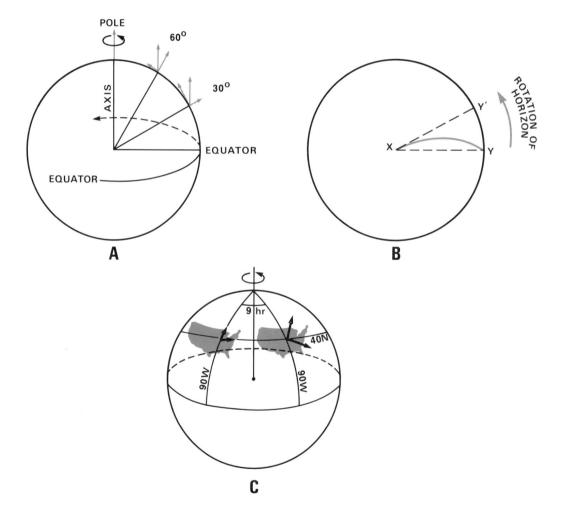

Force. Therefore, the deflection to the right increases, and the Coriolis Force swings around with the changing direction of the wind, remaining perpendicular to it at all points. Ultimately, the air moves with the Coriolis Force directed opposite to the pressure gradient. The wind then has a terminal velocity fast enough to cause the magnitude of the Coriolis Force to equal that of the pressure gradient force, and from then on there is no net force. Therefore, the wind neither accelerates nor decelerates, but moves with a constant speed and direction that is parallel to the contours, perpendicular to the pressure gradient. Thus, if the pressure gradient and Coriolis Force are the only two forces acting on the air, the wind does not move from high to low pressure but along lines of equal pressure, and thereby tends to perpetuate the original pressure pattern. This straight-line motion of air between the balanced pressure gradient and Coriolis Force is known as the *geostrophic wind.* It can be computed if one knows the pressure gradient and the latitude. The closer together the contours (the stronger the pressure gradient), the stronger the wind; and the lower the latitude, with the same pressure gradient, the stronger the wind.

Returning to the consideration of the atmospheric motion between equator and poles, because of the Coriolis Force neither the surface air, which tends to move from pole to equator nor the upper air, which tends to move from equator to pole, moves in a north-south direction but ends up moving in a zonal direction (parallel to the parallels of latitude). The pole-to-equator movement of surface air is deflected to the right in the Northern Hemisphere and the left in the Southern Hemisphere, which causes east-to-west winds at the surface in both hemispheres. The equator-to-pole-

Figure 5.2 (opposite page). The Coriolis Force. Part A illustrates that at any point on the earth's surface an object rotates about an imaginary axis parallel to the earth's axis of rotation. This rotation can be broken into two component parts, one that is parallel to the earth's surface and one that is perpendicular to it at that point. At the pole the entire rotation is parallel to the earth's surface around a vertical axis. Here the earth's grid rotates completely around once per day. At successively lower latitudes less and less of the rotation is parallel to the earth's surface around a vertical axis, and more and more is perpendicular to the earth's surface around a horizontal axis. At 30° latitude the horizontal rotation around the vertical axis is only one-half that at the pole. At the equator there is no horizontal rotation at all; an object moving across the surface would experience one complete rotation per day in a plane perpendicular to the earth's surface and would not change its direction with respect to the earth's grid. Part B illustrates the apparent deflection of a projectile shot from point X on the earth's surface in the Northern Hemisphere toward a target at point Y. During the flight of the projectile this portion of the earth's surface rotates counterclockwise around the observer at point X so that the target moves from point Y to point Y'. The projectile moves to point Y and thus misses the target. To an observer in space it appears that the projectile moved along a straight line from X to Y, but to the gunner at X, who is rotating with the earth, it appears that the projectile moved from X to Y along the curved solid line. Part C shows the rotation of the position of the United States during a nine-hour period. Assuming frictionless airflow and constant pressure gradient, air that initially was blowing from west or south at 40°N, 90°W will end up blowing from northwest or southwest because of the rotation of the continent under it.

LOW PRESSURE

HIGH PRESSURE

Figure 5.3. Frictionless airflow at 3100–3300 meters altitude quickly adjusts to the pressure pattern, so the wind blows with constant velocity parallel to the contours with the pressure gradient and Coriolis forces balanced against each other.

moving upper air ends up moving from west-to-east in both hemispheres.

Friction

All the foregoing considerations assume frictionless flow, and this is essentially true above the influences of the earth's surface, perhaps 300–600 meters in the air. But at the surface of the earth, friction between the moving air and the surface below becomes a significant force that must be taken into consideration. This changes the wind-force diagram to that shown in Figure 5.4. Friction (F) always acts opposite to the wind (W). This slows down the motion, which reduces the Coriolis Force, which allows the pressure gradient force (P) to overbalance the Coriolis Force (C) and pull the wind toward the pressure gradient. The Coriolis Force swings with the wind, remaining perpendicular to it at all times, and the friction swings with the wind, opposing it at all times. The motion soon adjusts to the three forces, so that the resultant of the Coriolis Force and friction is equal and opposite to the pressure gra-

dient. Thus, a balance of forces is again achieved, and the wind moves with constant speed and direction, crossing the isobars at some acute angle, commonly about 15°–30° over land and less than that over smoother water.

Hadley Cell

Thus, the surface air, which on a differentially heated earth might be expected to move from the cold poles toward the hot equator, ends up moving neither north-south nor east-west, but in some diagonal direction, from east-northeast to west-southwest in the Northern Hemisphere. There are now both zonal (latitudinal) and meridional (longitudinal) components. The upper air flow, which is essentially free of friction, nevertheless also takes on a meridional component from equator to pole in order to achieve mass continuity with the pole-to-equatorward component of the surface air. Although the zonal component is usually much larger than the meridional one, the meridional movement effects the latitudinal heat and momentum transfers

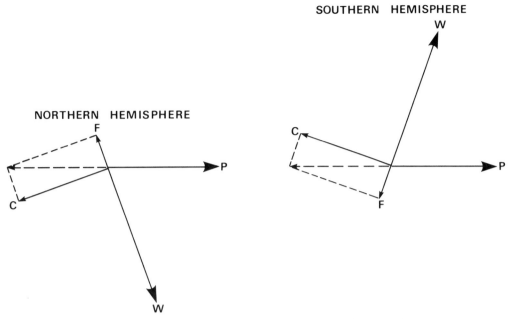

Figure 5.4. Wind-force diagrams of surface air flow involving friction. The same pressure gradient force (P) produces different wind directions (W) in the two hemispheres because of the reversal across the equator of the direction of the Coriolis Force (C).

necessary for the maintenance of the general circulation of the atmosphere and the heat balance of the earth. This meridional component comprises a vertical circulation cell that is known as the "Hadley cell," after the Englishman George Hadley who, in the eighteenth century, postulated these movements qualitatively to explain the trade winds (which will be explained in Chapter 12).

Momentum

Observation has indicated, and theory has demonstrated, that a simple Hadley cell cannot exist over an entire hemisphere. If the surface winds everywhere were easterly (from east to west), friction between the air and the earth's surface would soon slow the winds to no wind at all. In order to keep the trade winds blowing in low latitudes, there must be a mechanism for transferring the westerly (west to east) momen-

tum of the earth's rotation away from the easterly winds in low latitudes to the air at higher latitudes, and this transport of westerly momentum will then feed westerly winds at extratropical latitudes. Friction with the earth's surface would stall the winds whether they were all easterly or all westerly. There must be a balance between belts of alternating easterly and westerly surface winds, and there must be mechanisms for transferring westerly momentum from easterly wind belts to westerly wind belts.

Also, a simple meridional component of flow all the way from equator to pole in the middle troposphere is not feasible. This air, rising over the equator, is essentially stationary relative to the earth, and since the earth is rotating once a day, it has an average linear speed of more than 450 meters per second (m/s) (1000 miles per hour) from west to east. As the air moves poleward in essentially frictionless flow at upper levels, it comes into latitudes with constantly decreasing radii about the earth's axis of rotation. In order to maintain constant westerly momentum, the west-east velocity must increase correspondingly. By the time the air has moved half way to the pole, at 45° latitude, the radius of rotation has reduced to approximately 0.7 what it was at the equator, and the west-east absolute motion of the air must have speeded up to about 635 m/s (1400 mph). At the same time, the west-east speed of the earth's surface has decreased to a ratio of 0.7 what it was at the equator, and therefore the west-east air speed relative to the earth must be 330 m/s (737 mph)! Thus, the great west-east momentum imparted to the air by the rotation of the earth would cause inconceivably high wind speeds relative to the earth's surface at increasing latitudes. Such flow would become extremely turbulent and unstable, and zonal flow would break down into large whirls or eddies that would dissipate much of the energy.

Only at low latitudes, then, does a semblance of a Hadley cell exist. At higher latitudes momentum considerations become paramount, and the circulation of the atmosphere shows little or no direct connection with temperature patterns. The indirect "Ferrel" cell results in the middle latitudes from the transfer of westerly momentum from lower latitudes. It is related only indirectly to latitudinal temperature differences through momentum transfers effected by the Hadley cell of the lower latitudes, which is directly driven by heat differentials.

Resonance

The result is that, with the given energy inputs (latitudinal heat differentials and earth's rotation) and the character of the atmosphere (mass, vertical-temperature structure, and hydrostatic balance between gravity and pressure decrease with height),

the earth's atmosphere resonates in a given mode, much as a tuned instrument, that sets up circulatory systems that tend to be persistent in time and location. The air masses within circulating cells between resonating nodes tend to be somewhat self-contained, but there are slow exchanges of flow between them, effected largely by transitory circulatory systems on smaller spatial scales, that through their movements facilitate transfers of heat and momentum necessary to keep the general circulation going.

Cellular Flow

Most of the time the air does not flow along straight lines, but along curved lines around cells and vortices. For the wind to blow along curved isobars, another apparent force must be introduced. For example, suppose there is a circular cell of frictionless wind flow around a center of high pressure (Figure 5.5). A pressure gradient force (P) is directed outward in all directions from the center. Air will begin to move down the pressure gradients from the center toward the periphery of the cell. If this is in the Northern Hemisphere, the Coriolis Force (C) will act to the right and eventually cause the wind to flow clockwise parallel to the isobars at that point, with C equal and opposite to P. If the wind continues in a constant direction, however, it will go off on a tangent to the circular isobars and leave the circulation system. An additional force is necessary, directed toward the center of the high pressure cell, to cause a continual change in direction of the wind to keep it moving parallel to the curved isobars. In other words, the Coriolis Force must slightly overbalance the pressure gradient force. For this to happen, the wind must accelerate. Therefore, the wind flow around the curved isobars of a high pressure cell must be slightly stronger than the geostrophic wind with the same pressure gra-

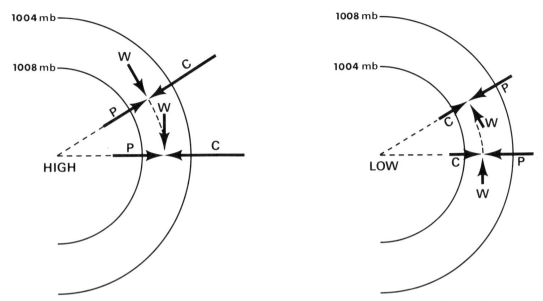

Figure 5.5. Gradient wind flows around high and low pressure cells in the Northern Hemisphere.

dient along straight isobars. The additional force thus produced is known as centripetal force, directed toward the center of the cell.

In the case of a low pressure cell, the pressure gradient is everywhere directed from the periphery toward the center (see Figure 5.5). The air begins to move from the periphery toward the center of the cell, but in the Northern Hemisphere the Coriolis Force acts to the right and makes the wind flow counterclockwise, parallel to the isobars. The centripetal force is again directed from the periphery toward the center, so that in the case of a low pressure cell the centripetal force and the pressure gradient force oppose the Coriolis Force. Therefore, around a low pressure cell the Coriolis Force is less than the pressure gradient force, and consequently the velocity of the wind must be less than it is around the high pressure cell or even the geostrophic wind along straight isobars.

Thus, for a given pressure gradient, the resultant wind will be fastest around a high pressure cell, slowest around a low pressure cell, and intermediate along straight isobars. Nonetheless, in nature the pressure gradient around low pressure cells is usually significantly greater than it is around high pressure cells, and it can be shown mathematically that the pressure gradient must weaken toward the center of a high pressure cell so that, ultimately, the wind becomes essentially calm at the center of a high pressure cell. In a low pressure cell, on the other hand, the pressure gradient normally increases toward the center of the cell, and high wind speeds may be achieved.

The balanced, frictionless flow just described around curved isobars is known as the *gradient wind*. Like the geostrophic wind, the gradient wind may be computed from the pressure gradient and the latitude. In the Southern Hemisphere, since the Coriolis Force is directed to the left of the motion, the air circulates around high and low pressure cells in opposite directions to those described for the Northern Hemisphere, counterclockwise around highs and clockwise around lows.

NORTHERN HEMISPHERE

SOUTHERN HEMISPHERE

EITHER HEMISPHERE

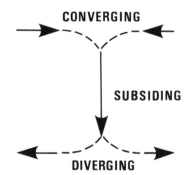

Figure 5.6. *Horizontal and vertical air flows in high and low pressure cells in the Northern and Southern Hemispheres.*

Near the surface of the earth, friction must be taken into account. This, again, will act opposite to the motion. It will reduce the velocity of the wind, thereby reducing the Coriolis Force and allowing the air to move gradually toward lower pressure. Therefore, in a high pressure cell, the air will gradually diverge from the central point in a spiraling motion outward toward the periphery in all directions. In a low pressure cell the air will spiral inward toward the center, thereby converging on the center from all points around the periphery (Figure 5.6).

In a high pressure cell, the constantly diverging air will tend to leave a partial vacuum at the center, and in a low pressure cell the converging air will cause an excess of air at the center. These imbalances must be compensated for by some additional air flow, and since these circulation systems are assumed to be at the surface of the earth, the only air flow that can compensate for deficits or excesses of air created by horizontal circulations must come from above. Therefore, in a high pressure cell a constant subsiding (downward) motion must take place in the middle of the high in order to replenish the air that is diverging from the center at the surface. In a low the excess air at the center must rise and diverge aloft. Therefore, the horizontal motion around highs and lows induces vertical motion, and although the magnitude of this vertical motion generally is much less than that of the horizontal motion, the vertical movement is the significant component that influences development of clouds and precipitation.

In summary, in the Northern Hemisphere atmospheric motion associated with high pressure cells is characterized by clockwise circulation, divergence, and subsidence. That associated with low pressure cells is characterized by counterclockwise circulation, convergence, and uplift. In the Southern Hemisphere the circulation directions are reversed, but the other motions remain the same: a high pressure cell is characterized by counterclockwise circulation, divergence, and subsidence, while a low pressure cell has clockwise circulation, convergence, and uplift (see Figure 5.6).

6 | OBSERVED ATMOSPHERIC AND OCEANIC CIRCULATION PATTERNS

Surface Atmospheric Flow

Schematic

The general features of the atmospheric circulation near the surface are essentially as shown in Figure 6.1. There are three pressure belts and three wind belts in each hemisphere. At or near the equator a continuous zone of low pressure relates to the greatest heat input over the course of the year, with resultant high average temperatures and reduced atmospheric density. Around 30° latitude in either hemisphere, a high pressure belt marks one of the resonating nodes of the atmosphere where excess air circulates into the zone in the upper troposphere and undergoes steady subsidence (settling motion). Around 60° latitude in both hemispheres another resonating node produces low pressure, and in the polar regions, generally low temperatures tend to produce dense air and shallow high pressure cells.

The surface winds flow between these pressure zones, responding to pressure gradients between high belts and low belts but being deflected by the earth's rotation, to the right in the Northern Hemisphere and to the left in the Southern Hemisphere. The broadest and most consistent of these wind belts are the trades, from the northeast in the Northern Hemisphere and from the southeast in the Southern Hemisphere, which blow between the subtropical highs and the equatorial low (or doldrums). The subtropical highs and equatorial low are the most consistent pressure belts on earth, so the trades are the most consistent wind belts. Hence their names, which relate to the commercial activities of the early sailing vessels that made use of these reliable winds, particularly between Europe and the New World. Rather than set out directly west from Europe to reach America, sailors traditionally moved parallel to the west coast of Europe and North Africa until they reached the trade winds and then headed west across the Atlantic. In so doing they had to cross the subtropical high belt (horse latitudes) with generally weak or even calm surface winds, where ships often became becalmed for days. Here was where the ship was becalmed in "*The Rime of the Ancient Mariner.*"

Between the subtropical highs and the subpolar lows the prevailing westerlies blow at the surface. In the Northern Hemisphere these are very inconstant because of the great amount of land at this latitude, which causes great land-sea differences in circulation, and the inconstancy of the subpolar lows, which generally are well formed only over the northern oceans during winter. Traveling circulation systems on a smaller scale cause the surface winds to shift frequently. In the Southern Hemisphere the westerlies are more constant, since there is practically no land at this latitude and the essentially circular, high plateau-like ice

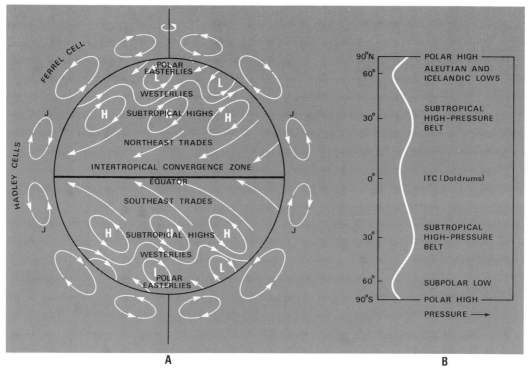

Figure 6.1. Schematic diagram of major pressure and wind patterns on the surface of the earth. Right-hand diagram has a cosine scale for latitude to represent correctly equivalent surface areas on the earth. This dramatically illustrates the dominance of the doldrums and subtropical high pressure belts in the global atmospheric circulation. And the left hand diagram correctly illustrates the great areal influence of the consequent trade winds and intertropical convergence zone. Idealized vertical circulations, including the Hadley and Ferrel cells, illustrate three cellular circulations in each hemisphere. The J's represent the subtropical jet streams at the tropopause level. (From Understanding Our Atmospheric Environment, *2nd ed., by Neiburger, M., Edinger, T.G., and Bonner, W.D., Copyright © 1982 by W.H. Freeman and Company. All rights reserved.)*

surface of Antarctica forms a good topographic barrier for the air to flow around. But even here, traveling circulation systems tend to break up the zonal circulation to some extent. Poleward of the subpolar lows the surface winds are commonly easterly, but this varies a great deal, particularly in the Northern Hemisphere where the Arctic Ocean is often not a region of consistently high pressure.

Mean Observed Patterns

Figures 6.2 and 6.3, which show observed sea-level pressures and prevailing surface

winds averaged over many years for January and July, reflect the major features of the schematic presented in Figure 6.1, plus additional complications produced by land-sea differences and seasonal changes in latitudinal positions and intensities. Here it can be seen that a low pressure belt in the vicinity of the equator forms a fairly continuous zone of convergence between the trade winds on either side at all times of the year, although it is distorted by monsoonal flows in Africa and Asia during summer. The subtropical highs are conspicuous throughout much of the year, particularly in the Southern Hemisphere where

Figure 6.2. Average sea-level pressures (in mb.) and prevailing surface winds, January. Width of arrow indicates consistency of wind direction. (From Fizikogeograficheskiy Atlas Mira [Physical-Geographical Atlas of the World] [Moscow, 1964], p. 40.)

Figure 6.3. Average sea-level pressures (in mb.) and prevailing surface winds, July. Width of arrow indicates consistency of wind direction. (From Physical-Geographical Atlas of the World, p. 41.)

there is less land to break up the zone. They are most prevalent over the oceans and usually are somewhat stronger and more extensive in area in summer than in winter. They are very well formed in the Northern Hemisphere over the oceans in summer where they expand and shift poleward to occupy much of the areas of the North Pacific and North Atlantic oceans. In winter, however, they become shrunken and weakened in intensity as they shift equatorward over the Northern Hemisphere oceans; and they are overshadowed by the apparent huge, intense high pressure cell that develops over Asia, which is related to the coldness of the landmass at that time of year. Both the Eurasian and North American landmasses tend to show pressure reversals between winter and summer because of the changing temperatures by seasons, the land being colder in winter and warmer in summer than the adjacent oceans. Since the Eurasian landmass is by far the larger one, this tendency is most pronounced over Eurasia.

The subtropical highs are dominant features in the troposphere, typically with great areal extent and great depth. The cells often extend upward to the middle troposphere and beyond, because they are warm-core highs whose pressures decrease with height less rapidly than their surroundings. They control the wind systems over broad areas on either side. Circulating clockwise in the Northern Hemisphere, they lead to the formation of the northeasterly trades blowing toward the equator on their southern flanks and the prevailing southwesterly winds blowing toward the subpolar lows on their northern sides. In the Southern Hemisphere, rotating counterclockwise, they originate the southeasterly trades blowing northwestward toward the equator and the northwesterly winds blowing southeastward toward the subpolar low.

The subpolar low is well formed in the Southern Hemisphere around the fringes of Antarctica. In the Northern Hemisphere it is limited primarily to the northern parts of the Atlantic and Pacific oceans during winter. Two intense lows at that time of year are centered in the vicinity of Iceland and the Aleutian Islands. They are known as the Icelandic Low and the Aleutian Low respectively. They spawn many traveling low pressure cells that sweep southeastward into Europe and North America to influence the weather over broad areas downstream throughout the middle latitudes of the Northern Hemisphere. During summer they are weak and diffuse, although there is still some semblance of the Low over Iceland and the northeastern part of the Canadian Archipelago. As can be seen in Figures 6.4 to 6.7, the postulated polar high appears to exist over Antarctica in the Southern Hemisphere but is illusory over the Arctic in the Northern Hemisphere. In both polar areas there is a greater tendency for high pressure over the pole in summer than in winter, but a rather nondescript pressure pattern occurs over the Arctic during all seasons. Therefore, the so-called polar easterlies in the Northern Hemisphere are not very discernable on most maps.

The Southern Hemisphere troposphere generally has a stronger zonal (west-east) circulation than the Northern Hemisphere, because the South Pole, which is more than three kilometers above sea level, is considerably colder than the North Pole, which is at sea level. Thus, the surface temperature gradient, which ultimately controls the strength of the atmospheric circulation in its respective hemisphere, is considerably greater between the equator and the South Pole than it is between the equator and the North Pole. This probably accounts for the average northward position of the intertropical convergence zone, between the trade winds, which generally lies north of the geographical equator (Figures 6.8 and 6.9). The Southern Hemisphere circulation, being stronger, tends to slop over the geographical equator so that the meteorological equator, on the average, lies to the north

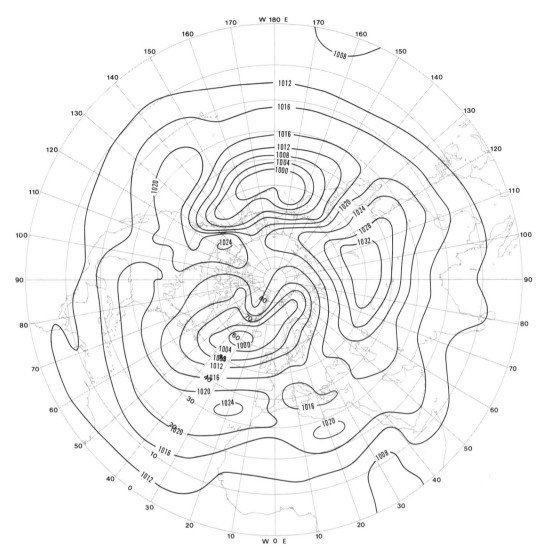

Figure 6.4. Mean sea-level pressure in mb., Northern Hemisphere, January. (From Selected Level Heights, Temperatures, and Dew Points for the Northern Hemisphere, *NAVAIR 50-1C-52.)*

of the geographical equator. During the Northern Hemisphere summer, when the Northern Hemisphere circulation is at its weakest, the convergence zone is pushed northward beyond the Tropic of Cancer in India, and in north Africa it is pushed northward to about 20° latitude (Figure 6.9). During January, on the other hand, the convergence zone lies only about 8°–10° south of the equator in the Indian Ocean

(see Figure 6.8). In South Africa it tends to orient north-south through the center of the continent. Over the Atlantic and Pacific oceans its position does not fluctuate as much, but on the average the convergence zone lies slightly north of the equator.

In both hemispheres the atmospheric circulation is stronger in middle latitudes and weaker in high latitudes during winter than during summer, because of seasonal

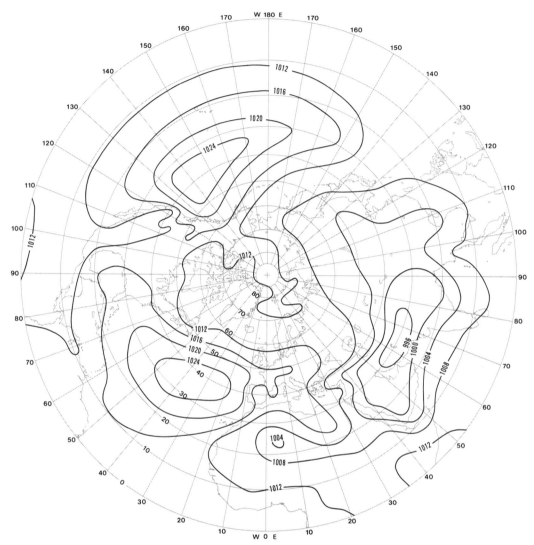

Figure 6.5. Mean sea-level pressure in mb., Northern Hemisphere, July. (From NAVAIR 50-1C-52.)

differences in latitudinal distributions of insolation (refer back to Figure 4.3). During summer the subtropical highs expand and move poleward at the expense of the westerlies, while during winter the westerlies expand their influences equatorward. The entire system of pressure and wind belts tends to shift poleward during summer and equatorward during winter, but not as much as the latitudinal shifts of insolation itself.

Causes of the Observed Pattern

Questions have naturally arisen as to why the earth's atmosphere near the surface circulates with three belts of winds in each hemisphere, rather than some other number. Experiments have been conducted in various ways (using both laboratory equipment and computer technology) to try to simulate the circulation of the atmosphere

Figure 6.6. Mean sea-level pressure in mb., Southern Hemisphere, January. (From Climate of the Upper Air, Part 1—Southern Hemisphere, *NAVAIR 50-1C-55.)*

and thereby ascertain its operative controls. One of the famous experiments of this kind was the so-called "dishpan experiment" carried out at the University of Chicago by David Fultz, who poured liquid into a dishpan-like vessel mounted on a rotating platform, which he heated at the rim and cooled at the center to simulate the equator and the pole. A substance was added to the liquid to make the circulation cells visible. Fultz found that different nodal patterns could be produced by varying the speed of rotation and/or the differential heating between the rim and the center of the pan. The faster the rotation or the less the heating differential, the greater the number of circulation nodes, and the slower the rotation or the greater the heating differential, the simpler the flow.

Thus, it appears that the earth's atmosphere circulates as it does because of the rate at which it rotates and the difference

Figure 6.7. Mean sea-level pressure in mb., Southern Hemisphere, July. (From NAVAIR 50-1C-55.)

in heating between the equator and poles. These, of course, are simply astronomical accidents, and if either of these inputs were changed the atmosphere might circulate quite differently. This is the primary reason why no one yet has dared to tamper with the Arctic ice cap. If it were melted, the temperature difference between the North Pole and the equator would be much reduced, and this might change the entire atmospheric circulation of the Northern Hemisphere, or perhaps of the entire earth. It probably would not simply cause a shift

in temperature belts with improved conditions along the Arctic coasts of America and Eurasia. It might even induce desert conditions or other detrimental effects over large sections of the earth's surface that now have tolerable climates for plant growth and human occupancy.

Ocean Currents

The same planetary controls that determine the general circulation of the atmosphere act on the water of the oceans, so it is logical to assume that oceanic circulations, at the surface at least, will mirror surface atmospheric circulations. To a great extent this is true (Figures 6.10 and 6.11). In the North Pacific and North Atlantic the general circulations are two large clockwise cells that correspond to the subtropical high pressure cells in the atmosphere above these regions. In the Southern Hemispheric oceans, large counterclockwise cells correspond to the subtropical highs above those regions. Near land masses, configurations of coastlines may set up localized swirls tangential to the main oceanic circulations; in higher latitudes, particularly around the fringes of the Arctic, currents may be induced by rapid changes in water temperatures.

Other factors enter into oceanic circulations beside density differences set up by temperature differences and Coriolis Force due to rotation of the earth. Currents may be caused by density differences due to salinity differences or turbidity (sediment content) differences along coasts. In addition, the atmospheric flow above the ocean and the surface of the ocean itself set up a frictional linkage. If the atmospheric flow is consistent, the water surface will take on a similar movement displaced somewhat to the right in the Northern Hemisphere and to the left in the Southern Hemisphere. Since the subtropical high pressure cells are consistent features of the atmosphere, under these cells the water movement is most reflective of the air movement.

Water is much denser than air and much more confined to basins, so it is not as free to react to density differences and the Coriolis Force as is the air. Nevertheless, surface waters, in particular, are free enough to move so that their patterns of circulation do resemble those of the atmosphere. Deeper currents may be quite different from the surface drifts, but they are of little climatic consequence, so this book will be concerned only with the surface currents of the oceans, primarily as they affect the climate along adjacent coasts.

Undoubtedly, the climatically most significant ocean current on earth is the Gulf Stream in the North Atlantic, which flows northeastward between Cuba and the southern tip of Florida and then parallels the eastern coast of North America almost to Newfoundland. It continues as the North Atlantic Drift northeastward past Britain and Norway, around the northern tip of Scandinavia, and into the Barents Sea of the Arctic north of European USSR. This large, warm current has a profound effect on much of Europe since there are few terrane barriers at these latitudes to stop the westerly winds that carry the marine air eastward. In the Norwegian Sea in winter, the Gulf Stream causes surface air temperatures to be as much as 26°C above the normal for the latitude. Much of Britain, the southern parts of which lie north of the Canadian border, experiences winters as mild as northern Florida, Georgia, and South Carolina, 15°–20° latitude farther south.

The counterpart to the Gulf Stream in the North Pacific is the Japan (Kuroshio) Current, which branches off the North Equatorial Current in the southwestern part of the North Pacific near the Philippine Islands and moves northward past Taiwan and Japan and then eastward across the Pacific as the North Pacific Drift. The Japan Current does not have as profound climatic effects on the eastern side of the Pacific as the Gulf Stream does on the eastern side of the Atlantic, because high mountains paralleling the west coast of North America do not allow much penetration of fresh Pacific marine air across the continent. Also, the North Pacific is separated from the Arctic during the winter by the narrow and frozen-over Bering Straits, which block any movement of surface water from the North Pacific into the Arctic during middle and late winter. This is not the case through the broad Norwegian Sea between the North Atlantic and the Arctic. Therefore, the climatic influences of the North Pacific Drift are not carried nearly as far poleward as

January

Figure 6.8. Streamlines and speeds of surface resultant winds, January. Solid lines with arrows (streamlines) represent net movement of air during January, and dashed lines (isotachs) represent resultant wind speeds in m/s (one m/s=2.237 mph). Speeds greater than 4.4 m/s (10 mph) are shaded. (Northern Hemisphere streamlines after W.M. Wendland and R.A. Bryson, "Northern Hemisphere Airstream Regions," Monthly Weather Review 109, no. 2 [Feb. 1981]. Isotachs and Southern Hemisphere streamlines from Y. Mintz and G. Dean, The Observed Mean Field of Motion of the Atmosphere, Geophysical Research Papers No. 17, Geophysics Research Directorate [Air Force Cambridge Research Center, 1952]. As shown in Understanding Our Atmospheric Environment, 2nd ed., by Neiburger, M., Edinger, T.G., and Bonner, W.D. Copyright © 1982 by W.H. Freeman and Company. All rights reserved.)

A

Figure 6.9. Streamlines and speeds of surface resultant winds, July. (See Figure 6.8 for symbols and sources.)

July

B

Figure 6.10. Surface ocean currents, January. Length of arrow represents consistency of direction, and width of arrow represents speed in knots (1.15 statute mph). Black arrows represent warm currents; white arrows, cold currents. Significant zones of convergence and divergence are indicated. (From Physical-Geographical Atlas of the World.)

Figure 6.11. Surface ocean currents, July. (From Physical-Geographical Atlas of the World.)

are the influences of the North Atlantic Drift.

Both the North Atlantic and North Pacific warm currents have branches moving southward along the eastern edges of their basins to complete the clockwise swirls of general oceanic circulations. These currents are cold relative to the latitudes into which they are flowing. (In all cases the designations "warm" or "cold" are relative terms.) Obviously, in most cases the California Current, flowing southward along the west coast of California and Baja California, has warmer temperatures, particularly in winter, than does the so-called "warm" Alaskan Current flowing northward into the Gulf of Alaska. But the California Current has a cooling effect on its region, while the Alaskan Current has a warming effect on its region. The counterpart of the California Current in the eastern North Atlantic is the Canaries Current which flows southwestward along the coasts of the Iberian Peninsula and northwestern Africa past the Canary Islands.

These equatorward-flowing cold currents continue westward across their respective oceans as relatively warm equatorial currents to complete the general clockwise circulation in each ocean, but these equatorial extensions have little climatic effect. It is interesting to note that in the Pacific Ocean during most of the year, both the North Equatorial Current and the South Equatorial Current lie to the north of the geographical equator, a reflection of an earlier statement regarding the more vigorous atmospheric circulation of the Southern Hemisphere. In between the two equatorial currents in the Pacific is an eastward flowing gravity current, the so-called "counter current," which serves to maintain mass continuity in the area, so the two equatorial currents do not pile up the water higher on the west side of the Pacific than on the east.

Warm currents are poorly developed in the Southern Hemisphere. Perhaps the most consistent one is the Brazil Current, which flows southwestward along the southeast coast of Brazil. But because of the triangular shape of eastern South America, much of the south Equatorial Current in the Atlantic, which should feed the Brazil Current, is shunted northwestward along the coast of Cape São Roque and moves north of the equator to join the current feeding the Gulf Stream. This leaves the Brazilian Current with much less flow than might be the case if the coastal shape of South America were different. Currents are localized and inconstant in the Indian Ocean, where they are greatly affected by the seasonal reversal of the monsoons during the year. In the South Pacific only a weak, diffuse warm current moves southwestward along the eastern coast of Australia.

The two currents with by far the most pronounced climatic influences in the Southern Hemisphere are large cold currents, the largest cold currents on earth. The most voluminous and climatically significant is the Peru (Humboldt) Current that flows northward along the west coast of South America from about 50°S latitude to almost the equator. In fact, it crosses the equator in the eastern Pacific shortly after diverging from the South American coast near the Gulf of Guayaquil. This has a profound effect on the climate along much of the west coast of South America, although its influences are limited to a narrow coastal strip because of the high Andes to the east. The counterpart of the Peru Current in the eastern South Atlantic is the Benguela Current that flows northward along the southwest coast of Africa.

All the cold oceanic currents paralleling west coasts of continents in both hemispheres are flowing in such directions that the Coriolis Force, plus frictional drag of winds, tend to throw them offshore. Within 100–200 kilometers of the coasts vertical circulations in the water are set up that bring cold water up from below to replace the surface water that is moving away from the coasts out to sea. It is obvious that in many cases the coldness of the surface water in these currents is due more to upwelling from below than to horizontal movements from higher latitudes, because cold pools of water are often found adjacent to land promontories, with warmer water on either side. For instance, during summer the water off the coast near San Francisco is often the coldest along the entire North American west coast. Obviously, the local factor of upwelling in that region is more significant than the southward horizontal movement of water from Alaska.

The four primary cold currents of the world—Peru, Benguela, California, and Canaries—are positioned under the equatorward eastern portions of the subtropical high pressure cells and further stabilize the atmosphere in those locations by cooling from below. Combined with the heating of atmospheric subsidence from above, they produce an extremely intense temperature inversion at or a few hundred meters above the surface. This limits upward movements of air to the shallow marine layer below the inversion base and induces a paradoxical climatic situation along adjacent coasts—little if any measurable precipitation, but consistently cool, moist surface air with frequent fog and overcast, low stratus clouds.

These four cold currents and the two large warm currents in the Northern Hemisphere are the major climatically significant ocean currents, but many other currents of more localized nature have pronounced climatic effects along their immediate shores. Examples are the cold Labrador Current flowing southward along the coasts of Labrador and Newfoundland in the North Atlantic, the Kamchatka and Kuril currents flowing southwestward along those coasts in the western North Pacific, the Somali Current along the eastern horn of Africa, the West Australian Current flowing northward along the western coast of Australia during much of the summer, and the cold Falkland Current flowing northward along the Patagonian coast through much of the year.

Upper Atmospheric Flow

The atmospheric circulation generally intensifies and simplifies upward through the troposphere. This is primarily due to the so-called "thermal wind effect," which is caused by different rates of decrease of air pressure with height because of different temperatures, and hence thicknesses, of air columns. For instance, the pressure decreases less rapidly with height over the warm equator than over the cold pole. Even if the pressures were the same at the equator and pole at sea level, the pressure at any given height over the equator would therefore be greater than the pressure at the same height over the pole. The higher into the troposphere, the greater the pressure difference (Figure 6.12). Thus, the wind increases from zero at the surface, with equal pressures at equator and pole, to increasingly greater speeds aloft with increasing pressure differences.

The difference in the winds at any two levels is known as the *thermal wind*, for it is dependent on the thickness of air columns between two pressure levels, which is a function of temperature. The thermal wind is a hypothetical flow along isotherms (lines of equal temperature) with low temperature to the left in the Northern Hemisphere and to the right in the Southern Hemisphere. With increasing height in the troposphere, the thermal wind component eventually becomes greater than any surface influence, so that by the time the middle troposphere (approximately the 500-mb. pressure level, or about 5000–6000 meters above sea level) is reached, the complex pattern of pressure cells and resultant winds at the surface has been replaced by one simple pressure gradient from equator to pole, with one large circumpolar whirl circulating from west to east, counterclockwise

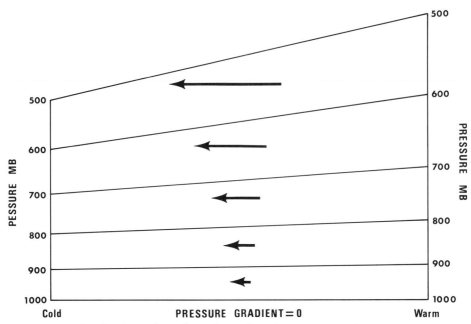

Figure 6.12. The thermal wind. At the surface the pressure is the same everywhere and the wind is calm. As the pressure decreases more rapidly with altitude in the cold air than in the warm air, the pressure gradient increases and the wind becomes stronger. Arrows represent relative magnitudes of the pressure gradient at increasing heights. The wind would blow into the paper (low pressure to the left) in the Northern Hemisphere and out of the paper (low pressure to the right) in the Southern Hemisphere.

in the Northern Hemisphere and clockwise in the Southern Hemisphere (Figures 6.13 to 6.16).

About the only surface features still evident at the 500-millibar level are the subtropical highs, which are displaced farther equatorward than they are at the surface. These high pressure cells, being composed of warm air, decrease in pressure only gradually with height, and therefore extend high into the troposphere. Since the pressure decreases with height most slowly on their warmer equatorward sides, the highs tilt upward toward the warm equatorward air. A cold high, on the other hand, such as that which forms over Asia during winter, is very shallow, since the pressure within it decreases rapidly with height. In fact, the high over Asia is no longer evident even at 850 millibars, only 1300–1500 meters above sea level (Figure 6.17). Low

pressures are just the opposite. Cold lows deepen aloft, since the pressure in the center of the low decreases with height faster than it does around the periphery. Warm lows, on the other hand, fill and disappear rapidly aloft, as the pressure decreases with height less rapidly in their centers than around their peripheries. Cold, intense lows tilt upward toward colder temperatures, generally to the northwest in North America or the northeast in Eurasia.

Wind speeds also increase with height through the troposphere, because air density decreases rapidly with height. With the same pressure gradient, the less the mass of air, the faster it will accelerate. Also, drags due to surface friction and turbulence are reduced with height, with resultant increased speeds.

The trend toward faster and more simplified air flow aloft generally continues

Figure 6.13. Mean 500 mb. chart, Northern Hemisphere, January. Contours are numbered in dekameters (tens of meters). (From NAVAIR 50-IC-52.)

upward through the troposphere to the tropopause, which is usually found between 200 millibars and 100 millibars (11,750–16,750 meters above sea level) during July over the Northern Hemisphere. Shortly above the tropopause, the circulation becomes confused as it adjusts to changed latitudinal temperature patterns. Since the tropopause is reached at lower levels in higher latitudes, the temperature increase with ascent in the stratosphere begins at lower levels at higher latitudes than it does at lower latitudes, and eventually a temperature reversal takes place, with warmer temperatures over high latitudes and colder temperatures over low latitudes. The thermal wind then changes the mid-stratospheric circulation to east-to-west (Figure 6.18).

Such a reversal with height does not take place during winter, when the polar area is constantly in the dark and little ultra-

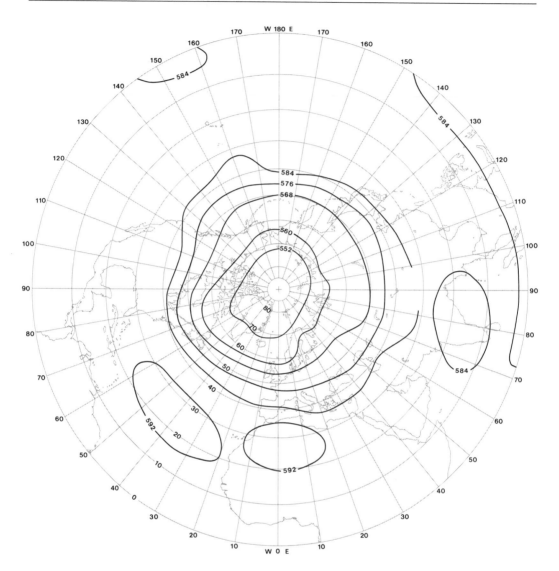

Figure 6.14. Mean 500 mb. chart, Northern Hemisphere, July. Contours in dekameters. (From NAVAIR 50-IC-52.)

violet light is absorbed at the level of the stratopause. The stratosphere at this time of year remains colder at every height over the pole than over the equator. Thus, during winter westerly winds extend upward to the outer limit of the atmosphere over the entire hemisphere, except for the equatorial region itself.

Figure 6.18 may give the impression that the westerlies, which expand latitudinally aloft to at least the tropopause level in all seasons, dominate the circulation of the atmosphere. It must be remembered, however, that most of the mass of the atmosphere is at lower altitudes, and at the surface of the earth easterly winds occupy more area than westerly winds do. Also remember that the earth is much larger around at the equator than at the poles and that half the earth's surface lies between

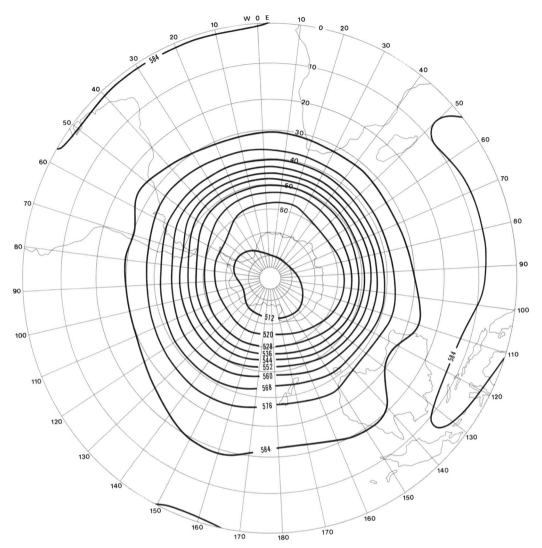

Figure 6.15. Mean 500 mb. chart, Southern Hemisphere, January. Contours in dekameters. (From NAVAIR 50-IC-55.)

30°N and 30°S latitude. At all times the easterly trades span this latitudinal zone, and many times overlap it to 35° latitude or more. And although the polar easterlies are very shallow, they do generally prevail at the surface from the poles equatorward to about 65° or 70° latitude. Therefore, a considerably greater portion of the earth's surface is affected by easterly winds than by westerly. The east-west movement of this denser air near the surface counter-balances the greater volume of westerlies in the middle and upper troposphere and above.

Much still remains to be learned and explained about the circulation of the high atmosphere. For instance, in the equatorial stratosphere there appears to be a 2.2-year cycle of alternating layers of easterly and westerly components, which descend through the stratosphere at the rate of about one kilometer per month. This 26-month

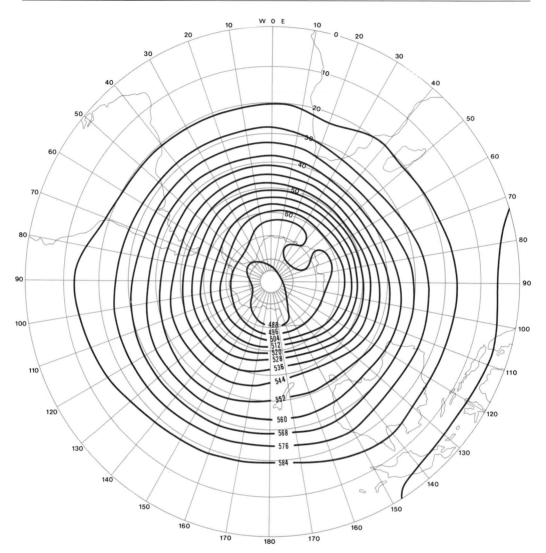

Figure 6.16. Mean 500 mb. chart, Southern Hemisphere, July. Contours in dekameters. (From NAVAIR 50-IC-55.)

cycle, of course, does not correspond to any astronomical period such as the year, and therefore piques much curiosity.

Figure 6.19 attempts to demonstrate the vertical flow of the atmosphere at different latitudes. It illustrates that between the equator and 30° latitude the dominant feature in the troposphere is a Hadley cell with rising air currents over the equator and descending air currents in subtropical zones. These vertical movements initiate equatorward flow at the surface and pole-

ward flow in the upper troposphere. Since the troposphere is warm over the equator, it extends to very great heights and is considerably higher in the tropics than farther poleward. In fact, the tropopause typically is broken at two latitudes, one near the descending poleward arm of the Hadley cell and the other in the subpolar zones around 60° latitude.

The mid-latitude circulation is more complicated, being characterized by traveling waves and vortices that ultimately

Figure 6.17. Mean 850 mb. chart, Northern Hemisphere, January. Contours in dekameters. (From NAVAIR 50-IC-52.)

transport heat and westerly momentum from the poleward edges of the Hadley cells to drive the westerly Ferrel cells of the middle latitudes. In the vertical, there is no complete circulation; westerly winds generally prevail at all altitudes throughout the troposphere. But averaged over time, there is a gradual movement upward of warm moist air from lower latitudes and a subsiding of polar air from higher latitudes along a baroclinic zone that slopes equatorward from the top of the troposphere to the earth's surface.* (This so-called *polar front* will be discussed in detail in Chapters 9 and 10.) North of 60° latitude there is little vertical movement in the

---

*An unstable zone in which air pressure surfaces do not coincide with air density surfaces.

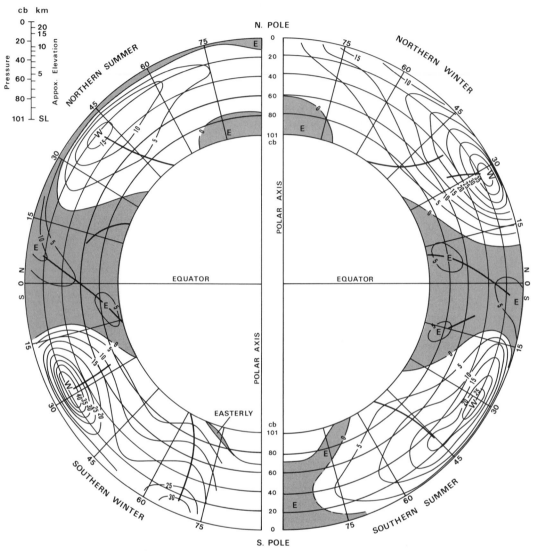

Figure 6.18. Mean zonal (west-east component) winds around the earth, by latitude and altitude. Isotachs (lines of equal air speed) are labeled in meters per second. (After Y. Mintz, "The Observed Zonal Circulation of the Atmosphere," Bulletin, American Meteorological Society *[1954]:209.*

troposphere. The air there is dominated by horizontal mixing that generally is not as vigorous as in the middle latitudes.

Jet Streams

At the subtropical and subpolar breaks in the tropopause, large horizontal temperature gradients set up high-speed flows of air that have become known as the jet streams. According to the World Meteorological Organization, any speed exceeding 30 meters per second (67 mph or 108 kilometers per hour) may be called a jet stream. Speeds vary along the axis of a jet stream, particularly along the polar front jet. Strongest jet streams have been recorded during winter over southern Japan, where speeds up to 140 meters per second (500 kilometers per hour, or 310 mph—270 knots)

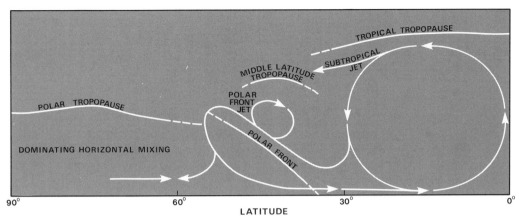

Figure 6.19. Schematic representation of the meridionic circulation and associated jet cores. (After E. Palmén, "The Role of Atmospheric Disturbances in the General Circulation, "Quarterly Journal of the Royal Meteorological Society, Vol. 77 [1951], pp. 337–54.)

have been observed. Secondary maxima generally occur off the east coast of North America and in the central part of North Africa (Figure 6.20). Along the edges of jet streams, wind sheers produce sharp turbulence as small vortices form between the fast moving air streams and the more slow-moving air around them. This accounts for the rough air encountered by aircraft upon entering or leaving jet streams.

Two jet streams, the subtropical jet and the polar front jet, are typical during winter in the Northern Hemisphere. The subtropical jet is located, on the average, at about 28°N latitude at about the 200-millibar level, between the tropical tropopause (at approximately 100 mb) and the mid-latitude tropopause (at approximately 250 mb). The polar front jet lies near the 300-millibar level between the mid-latitude tropopause and the polar tropopause. The polar front jet meanders and shifts its position much more than the subtropical jet, so that, while it is usually the stronger of the two, it tends to become obscured when averaged with all jet stream occurrences over the period of a month or more. Thus, most representations of the average position of the wintertime jet in the Northern Hemisphere approximate the position of the subtropical jet centered over the subtropical high pressure zone at the surface (Figure 6.20). But

on daily maps during winter, most of the time the two jets are discernible, and the highest wind speeds are usually encountered at higher latitudes, averaging about 50°N (Figure 6.21). During summer in the Northern Hemisphere, the subtropical jet tends to disappear, and the average position of jet streams in the Northern Hemisphere shifts poleward (Figure 6.22).

Although observations are scarce in the Southern Hemisphere, the jet streams there appear to be similar to those in the north, with perhaps less fluctuation in position and intensity than in the north because of smaller land masses in the Southern Hemisphere. The strongest westerly winds are found south of Africa and over the Indian Ocean. During winter a separate belt of subtropical jets is found near 30°S, which is most pronounced over eastern Australia. This subtropical jet apparently disappears during the Southern Hemisphere summer. All jet streams tend to be weaker in summer than in winter, as is the entire circulation of the troposphere, since the latitudinal temperature gradient is reduced between equator and pole during summer.

Standing Waves

The circumpolar west-to-east circulation of the middle and upper troposphere is not

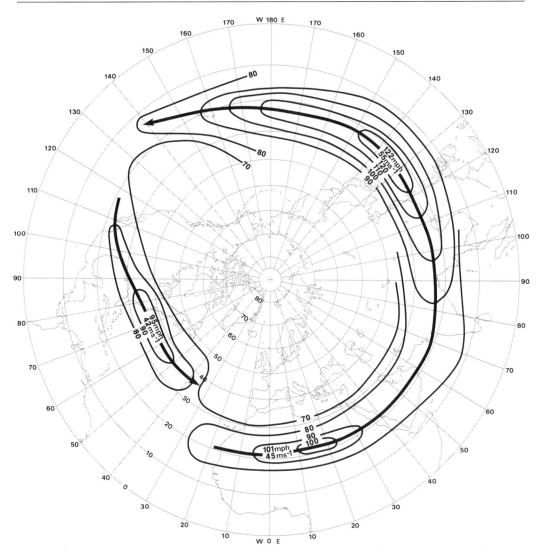

Figure 6.20. Average positions and strengths of jet streams in the Northern Hemisphere, January. (After Namias and Clapp.)

exactly circular, but meanders through equatorward-extending troughs and poleward-extending ridges, some of which are quasi-stationary in space over protracted periods of time. These are the so-called *standing waves* of the upper troposphere or the "Rossby waves," named after Carl Rossby who, in 1938, first described them adequately and dealt with them quantitatively. These waves are most pronounced in winter. They generally undergo sequences of events that increase their meridional magnitudes until they become so elongated that they become cut off. Then isolated pools of cold air remain for a while in lower latitudes, and warm pools remain in higher latitudes, while the west-east flow reestablishes itself with only minor undulations to begin the sequence all over again (Figure 6.23).

As can be seen on Figure 6.13, the January average flow shows two very pronounced troughs, one over the eastern Canadian Archipelago and another across east-

Figure 6.21. 300 mb. chart, 6 February 1981. Isotachs (lines of equal air speed) are black lines numbered in knots (1.15 statute mph). Regions with air speeds greater than 70 knots are shaded. Gray lines are contours in dekameters. Segments of both the polar and tropical jets are discernible particularly over the central Pacific and northwestern Canada, the northern and central Atlantic, and northern Europe and north Africa.

ern Siberia to the Sea of Okhotsk, and a lesser trough through eastern Europe. Relatively pronounced ridges appear over the eastern North Pacific, extending northward into the Bering Strait, and over the eastern North Atlantic, culminating over central Greenland, while a very weak, broad ridge extends over much of the central part of

Asia, particularly northward in Siberia. In the upper reaches of the troposphere the jet streams follow these sinuous routes of the standing waves around the hemisphere.

As with the jet streams, monthly averages tend to obscure some features that are apparent on daily maps. While mean monthly 500-millibar maps show an average

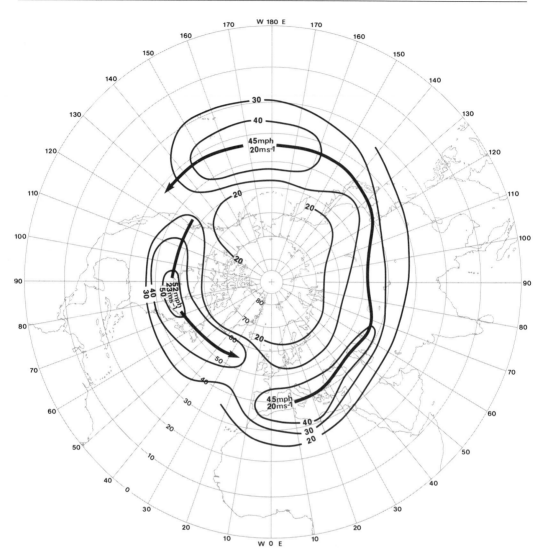

Figure 6.22. Average positions and strengths of jet streams in the Northern Hemisphere, July. (After Namias and Clapp.)

of three standing waves around the Northern Hemisphere, most daily maps show five, some less, some more. The 500-millibar map for 12 April 1983 illustrates a rather extreme case of eight fairly distinguishable troughs around 50°N latitude, plus some short waves in between (Figure 6.24).

The origin of these standing waves is undoubtedly tied to topography and heat sources and sinks, such as the thermal differences between sea and land. Perhaps the greatest influence on the Northern Hemisphere flow is the Rocky Mountains in western North America, which lie essentially perpendicular to the zonal flow and probably cause the deep average monthly trough over eastern North America. The weak average trough over eastern Europe may be simply a resonance wave, a downstream ripple of the anchor wave over eastern North America. But the tem-

Figure 6.23. Schematic representation of a wave sequence (index cycle) in the upper troposphere.

perature differences between sea surfaces and land surfaces in coastal regions may be just as important as topographic barriers. For instance, whenever sea surface temperatures are high in the eastern Pacific, a pronounced ridge forms in the upper troposphere over the eastern Pacific and western North America, which has profound effects on the weather downstream across most of North America.

In the Southern Hemisphere, probably the greatest influence is the Andes mountain chain in western South America. Nonetheless, since there is much less land in the higher middle latitudes in the Southern Hemisphere, the air flow there seems to be much more zonal (west-east). This apparent simplicity may be partially due to lack of data in the Southern Hemisphere. Plots of balloon paths indicate the presence of five waves in the Southern Hemisphere, similar to the daily situation in the Northern Hemisphere (Figures 6.25).

The standing waves of the middle and upper troposphere remain quasi-stationary, while shorter waves and vortices below move along their contours. It may take weeks or even months for the long waves to shift their positions significantly, and thus they become conservative factors in the atmospheric flow and allow for some extended forecasts that could not be predicted from the rapidly fluctuating flow patterns of the surface air. Although the

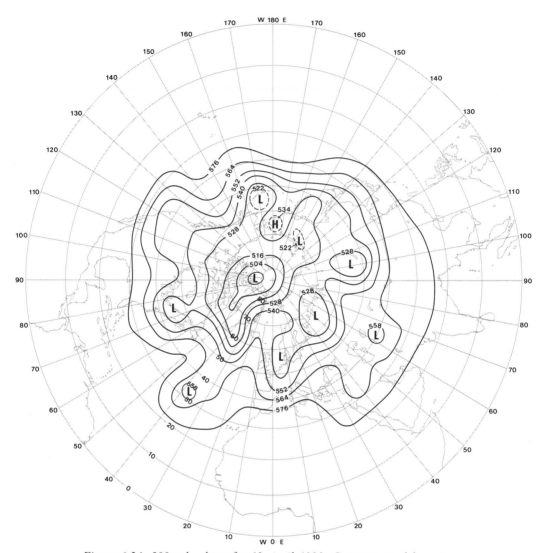

Figure 6.24. 500 mb. chart for 12 April 1983. Contours in dekameters.

standing waves of the upper troposphere tend to remain in semiconstant forms over prolonged periods of time, they sometimes undergo rapid transitions into different formations, which then bring about abrupt shifts in general weather conditions over broad sections of the earth's surface. A region that might have experienced a prolonged rainy spell of several weeks' duration may next find itself in a prolonged spell of droughty conditions. Or a similar contrast of temperatures might take place. The standing waves can slowly shift their po-

sitions either downstream, with the wind, or upstream, against the wind. The direction of motion largely relates to the length of the wave—the longer waves tending to regress upstream, while the shorter waves move downstream. Usually, shorter waves move faster than long waves.

Vorticity

The meandering of the upper air streams is a natural phenomenon that relates to the spin of air particles across the face of the

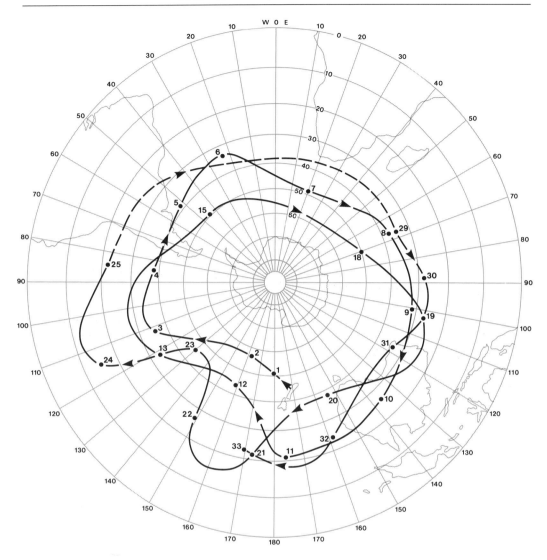

Figure 6.25. Path and daily positions of a constant-level balloon launched from Christ-church, New Zealand, 30 March 1966. The balloon drifted for 33 days at an altitude of about 12 kilometers, during which it circled the Southern Hemisphere three times. (From National Center for Atmospheric Research.)

earth. The amount of spin per unit area is known as *vorticity*. This has two components, one relating to the Coriolis Force, which increases with latitude, and one relating to the curvature of the flow itself, known as *relative vorticity. Positive relative vorticity* is defined as increasing cyclonic flow (counterclockwise in the Northern Hemisphere and clockwise in the Southern Hemisphere), and *negative vorticity* is de-fined as decreasing cyclonic flow or increasing anticyclonic flow (clockwise in the Northern Hemisphere and counterclockwise in the Southern Hemisphere). If an air stream is moving with some meridional component, say from southwest to northeast in the Northern Hemisphere, it will be moving into higher latitudes, which will increase the Coriolis Force, which increases that component of vorticity. For the ab-

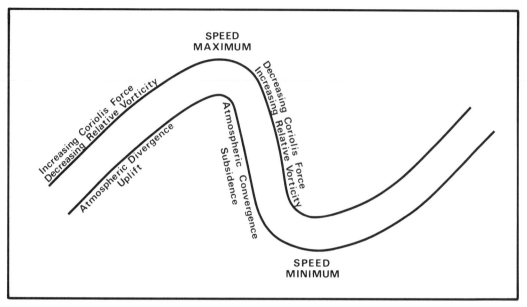

Figure 6.26. Schematic representation of changing Coriolis Force and relative vorticity in flow along a standing wave.

solute vorticity (the total of the two components) to remain constant, the relative vorticity due to curvature must decrease. Thus, the northeasterly moving flow must become more anticyclonic, or curve eastward and ultimately southeastward (Figure 6.26). As the flow starts to move southward again, the vorticity component due to the Coriolis Force begins to decrease, and therefore the relative vorticity due to curvature must increase once more, thus turning the flow in a cyclonic (counterclockwise) direction, which ultimately sends the air back to the northeast again. Thus, the air moves downstream in a wavy pattern with the relative vorticity constantly adjusting to the changing latitude, and the meandering flow tends to perpetuate itself.

Divergence and Convergence

The standing waves of the upper troposphere are important not only for steering, and thus predicting, the future motion of storms and other more localized features in the lower troposphere, but also because

they aid in predicting the development of such features; i.e., whether they are going to strengthen or weaken. Since for a given pressure gradient the wind moves faster in anticyclonic flow (around a high pressure ridge) than in cyclonic flow (around a low pressure trough), the air, in flowing along a meandering wave pattern, will be constantly changing its wind speed. This will produce speed convergences and divergences in the flow that will induce vertical motion in the air.

For instance, in Figure 6.26, the air moving from southwest to northeast and undergoing increasing anticyclonic curvature will tend to speed up, while the air moving from northwest to southeast and coming into influences of increased cyclonic curvature will tend to slow down. Therefore, the air will experience a divergence or pulling apart in the horizontal plane in the southwest-to-northeast portion of the wave, and a piling up or convergence in the northwest-to-southeast portion of the wave.

The air deficit in the region of divergence

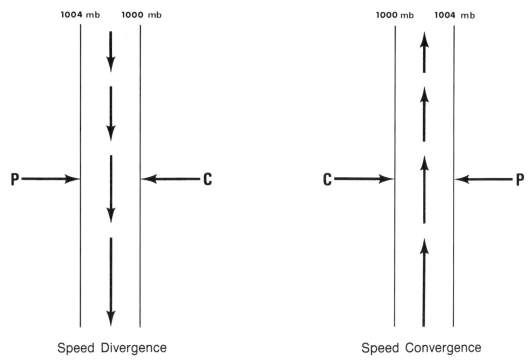

1004 mb   1000 mb   1000 mb   1004 mb

P ——→   ←—— C   C ——→   ←—— P

Speed Divergence   Speed Convergence

Figure 6.27. Speed divergence and convergence induced by latitudinal changes in Coriolis Force with north-south flow in a field of constant pressure gradient, Northern Hemisphere.

will have to be compensated for by vertical motion. And since this is in the middle troposphere, air can move into the layer from both above and below. Since most of the storm development and cloud formation is below this level, the important movement is the vertical motion rising from below. This will tend to strengthen storm development, which will induce cloud formation and possible precipitation. (These processes will be discussed more fully after the topic of moisture has been treated.) The portion of the wave that is experiencing convergence will have an excess of air that must move vertically downward below and upward above. Again, it is the downward motion below that becomes important to the consideration of weather conditions, and this subsiding air would weaken any storm moving along this poriton of the track and tend to produce clearing skies and fair weather.

Thus, other things being equal, the weather conditions in opposite portions of standing waves tend to be opposite in character: increasing storminess, cloudiness, and precipitation in southwest-northeast-oriented eastern branches of troughs or western branches of ridges, and clearer, drier weather in northwest-southeast-oriented western branches of troughs or eastern branches of ridges in the Northern Hemisphere. Since the standing waves do shift slowly with time, over distances of several hundred or even several thousand kilometers, an area might experience a prolonged period of clouds and rain as the wave occupies one position, and then experience a prolonged period of drought as the wave shifts to another position.

Since essentially opposite types of weather can be expected simultaneously in places about one-half wavelength apart, the Soviets, with their large west-east territory, made use of this knowledge during the

1950s to develop the so-called "Virgin Lands" east of the Urals, which are essentially one-half wavelength away from the older farming regions in the Ukraine. These two areas, which are both droughty, seldom experience extreme drought simultaneously. When it is exceptionally dry in the Ukraine, it is usually exceptionally moist east of the Urals in western Siberia and northern Kazakhstan, and vice versa. By planting both areas with similar crops, the Soviets can assure themselves of a fairly stable year-to-year overall production.

Since cyclones tend to intensify on poleward-moving limbs of waves, and since low pressures tilt toward colder temperatures, the traveling low pressure vortices of the middle latitudes eventually end up moving poleward and ultimately merge in the general region of the subpolar lows to cause constant regeneration of those systems. Conversely, anticyclones intensify on equatorward-moving limbs of waves and tilt toward warmer temperatures, so high pressure cells ultimately move equatorward to merge with the subtropical high pressure belts and constantly regenerate those systems.

Horizontal convergence or divergence can be produced either by directional movements that come together or spread apart (see Figures 6.8 and 6.9) or by speed variations within flows, either of which can be achieved in either straight-line flow or curvilinear flow. Often the speed variations within an air stream are brought about by changing Coriolis Force with changing latitude. For instance, assume straight north-south-oriented isobars in the Northern Hemisphere, as shown in Figure 6.27. In the left-hand diagram the pressure gradient force (P) is directed toward the east and is constant all along the isobars, since they remain equally spaced. Assuming frictionless flow, the wind will flow from north to south with the Coriolis Force (C) directed westward equal in magnitude to the pressure gradient force. As the air moves into lower and lower latitudes, the Coriolis Force tends to decrease because of the decreasing latitude, and this allows the wind to be pulled partially toward the pressure gradient. If the air motion has a component directed down the pressure gradient, it will accelerate. This will increase the speed of the wind, which will increase the Coriolis Force, which will pull the wind back to the original direction, and again the Coriolis Force will balance the pressure gradient force. The net result is an increase in wind speed but no change in direction. Hence, in an equatorward-flowing air stream, horizontal divergence will occur that will induce vertical motion toward that air stream. If the air is at the ground and this is a surface wind, the only direction the air can come from is above, and therefore the horizontal divergence induces subsidence which causes compression, heating, and fair weather. Conversely, air moving from equator to pole along straight, equally spaced isobars will constantly slow down, and air from behind will tend to pile up on air in front, which produces horizontal convergence, which at the surface would produce uplift (right-hand diagram).

It is interesting to note that with the same pressure gradient, the wind will have a higher velocity at low latitudes than at high latitudes. At low latitudes, because of reduced Coriolis Force, the wind must reach higher speed in order for the Coriolis Force to be large enough to balance the pressure gradient force. In reality, however, pressure gradients usually are not constant over extended latitudinal distances, but diminish toward lower latitudes, so that in most cases the winds are quite weak in low latitudes and there are few isobars on pressure maps in equatorial regions. Since the Coriolis Force reduces to zero at the equator, the concept of geostrophic wind breaks down in equatorial areas.

In the middle troposphere, as stated a

few paragraphs back, the same convergence and divergence would produce subsidence and uplift respective of the air below the air stream. At the top of the troposphere, if segments of jet streams are present, strong divergence is most frequently observed in the left-front quadrant of a jet core, followed by the right-rear quadrant. Thus, most uplift, and hence cloud formation and precipitation, will occur to the right of the axis where air enters a jet core and to the left of the axis where air leaves a jet core.

7 | ATMOSPHERIC MOISTURE

Moisture is the third basic ingredient, along with heat and motion, that determines the weather over the earth. It not only supplies the water for cloud formation and precipitation, but also, through changes of state, is intimately involved with great quantities of energy in the form of latent heat. All the moisture in the atmosphere is derived from the earth's surface through the process of evaporation. A great deal of heat is required to change the state of water from liquid to gas. This is known as *latent* (hidden) heat because it does not effect a change in temperature. This heat is not lost forever, but is released again once the water changes back from gas to liquid. The latent heat released may then be utilized to effect a temperature change in the air, or for some other purpose. Also, latent heat is involved in the change of state from solid to liquid and vice versa.

Latent Heat

To envision the quantity of heat involved in the change of state of water, recall that the heat unit, the calorie, is defined as the heat that is required to raise the temperature of 1 gram of water 1°C. To raise the temperature of 1 gram of liquid water from the freezing point, 0°C, to the boiling point, 100°C, would require 100 calories of heat. But to change that 1 gram of water at the boiling point to water vapor at the boiling point would require 540 calories of heat. Also, if water at the freezing point were

initially in the frozen state, it would take 80 calories per gram to melt the ice into water at 0°C. Therefore, it can be seen that it requires almost as much heat to melt ice as it does to raise the temperature of water from freezing to boiling, and it requires 5.4 times as much heat to evaporate water at the boiling point as it does to raise its temperature from freezing to boiling (Figure 7.1).

Thus, evaporation and condensation involve truly great quantities of heat. Under natural conditions evaporation usually takes place much below the boiling point of water. This requires more heat. At 0°C, the latent heat of evaporation is 596 calories per gram. Sometimes there is a direct change from solid to gas, as in the case of evaporation of snow. This is known as *sublimation*. The latent heat of sublimation is the sum of the latent heat of fusion (or melting) and the latent heat of evaporation (or condensation), or a total of 676 calories per gram if the sublimation takes place at 0°C. If ice at 0°C is melted and then the liquid water is heated to the boiling point and evaporated, the total amount of heat necessary would amount to $80 + 100 + 540 = 720$ calories.

Water Vapor

The water vapor in the atmosphere can be expressed in a variety of ways. All are significant, since some serve some purposes and others serve other purposes. *Absolute humidity* is defined as mass of water vapor per volume of air. This is not a conservative number, since as the air moves about it expands and contracts. Thus, the absolute humidity can change drastically without any addition or subtraction of moisture.

A more conservative measure is *specific humidity,* which is defined as the mass of water vapor per mass of air. A closely related value is the *mixing ratio,* which is the mass of the water vapor compared to the mass of the dry air. Since air is always much more abundant than water vapor, there is little difference between *specific humidity* (which is the mass of water vapor divided by the total mass of the air including the water vapor) and the mixing ratio (the ratio of the water vapor to the dry air). These values are quite useful in technical work, such as weather forecasting, as they are measures of such things as latent heat energy available in the air due to its water vapor content, availability of precipitable water, atmospheric stability characteristics, and so forth.

To the layman, however, the fact that the air at a given time might contain 10 grams of water vapor per kilogram of dry air probably holds little meaning. More meaningful might be the degree of saturation of the air, which noticeably affects the physiological workings of the human body and all other living organisms, as well as cloud formation, precipitation processes, and so forth. This expression, known as *relative humidity,* is the ratio of the actual humidity of the air compared to the saturation humidity of the air, both expressed in the same units. Thus, if the water vapor content of the air has been measured to be 7 grams of water vapor per kilogram of dry air, and the saturation mixing ratio of the air is known to be 10 grams per kilogram, the relative humidity is 70 percent.

The saturation mixing ratio of the air depends on temperature; it increases at an increasing rate with increasing temperature

Figure 7.1. Heat requirements to change the state of water.

(Figure 7.2). Thus, a specific change in temperature will effect a greater change in saturation mixing ratio at higher temperatures than at lower temperatures. This fact will become very important when stability characteristics of the air are considered in the next chapter.

Relative humidity can be changed either by adding moisture to or subtracting moisture from the air, or by changing the air temperature. Temperature changes normally take place more quickly than moisture changes, and so short-term relative humidity changes are usually related primarily to temperature changes. For instance, relative humidity is usually rather high early in the morning when the temperature is near its minimum point, but drops as the air heats up during the morning. Thus, in Figure 7.2, if the air on a given morning is saturated at 10°C and warms up during the morning to 20°C without any addition or subtraction of moisture, its relative humidity would have dropped to little more than 50 percent. Conversely, the relative humidity tends to increase during the night as the temperature decreases.

If the air contains much water vapor and the night is clear and still so that much heat is radiated outward from the earth's surface, the temperature drop may be enough to reduce the capacity of the air to the point where it can barely hold all the moisture that it has in it. The air then becomes saturated, and if cooling continues, some of the moisture must be condensed into liquid droplets and may settle on surface features in the form of dew. Thus, the temperature at which the air becomes saturated is referred to as the *dew point*, which is another measure of the amount of moisture in the air. Although the dew point is expressed as a temperature, it is a measure of the actual amount of moisture in the air. This is the expression commonly used in weather reports. With the temperature it gives a rough indication of the degree of saturation of the air and the temperature at which the air will become saturated if cooling takes place.

For certain purposes it becomes most useful to express the water vapor content of the air in terms of the pressure it exerts. Since the air is simply a mixture of gases, its total pressure can be divided into its component parts: that due to nitrogen, that due to oxygen, and so forth, and that due to water vapor. It is the *vapor pressure* that determines the movement of moisture across an interface, such as a water surface in contact with air above it. If the vapor pressure of the water surface is greater than the saturation vapor pressure of the air above it, moisture will move upward from the water surface into the air, whether the air is saturated or not. This phenomenon is often observed over relatively warm water bodies, such as lakes and rivers, early in winter before they freeze. If cold air moves across the warmer water surface, moisture will be injected upward into the cold air and will probably supersaturate the cold air, which will cause recondensation that produces what is known as steam fog.

The evaporation rate from a given surface depends upon the vapor pressure difference between the evaporating surface and the medium into which the moisture is moving. Vapor pressure difference also accounts for moisture movement from small water droplets to large water droplets and from a water surface to an ice surface at the same temperature, since the vapor pressure at the surface of a water droplet is inversely proportional to the radius of the droplet, and it is greater over a water surface than over an ice surface at the same temperature. These facts will be taken up again in Chapter 13 in a discussion of cloud formation and precipitation. At the same time it will be stressed that the air not only contains water vapor but also water in liquid and solid states in droplets that are small enough to be suspended in the air by the motion of the air. These liquid and solid droplets are visible in clouds and fog; water vapor cannot be seen. Thus, most of the time not only an exchange of moisture is taking place at the atmosphere-earth interface but also changes of state of moisture from solid to liquid to gas, and vice versa, are constantly taking place within the atmosphere. Water is unique among the atmospheric gases in that it occurs in all three states within the normal range of temperature within the atmosphere.

Since all moisture is derived from the earth's surface through the process of evaporation, the moisture content of the air is much greater in the lower troposphere than at higher altitudes. Little is found above the tropopause, since vertical mixing is generally limited above that level. Within the troposphere may be found distinct layers, produced by movements of air, some being relatively dry and others containing much moisture.

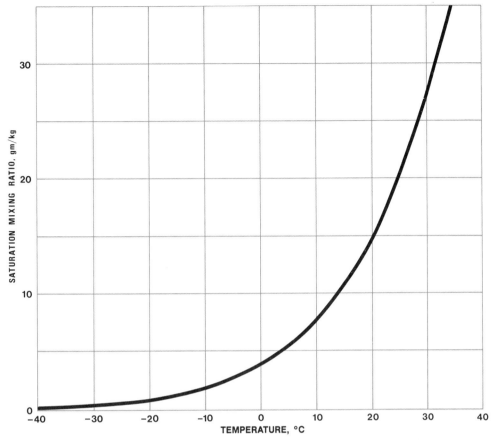

Figure 7.2. The saturation mixing ratio increases at an increasing rate with increasing temperature.

Geographical Distribution of Evaporation and Atmospheric Humidity

The amount of evaporation that takes place over a surface depends upon the availability of moisture at the surface and the vapor pressure gradient between the surface and the air above it. Since plenty of moisture is always available at the surface of the oceans, in general there is more evaporation over the oceans than over the continents at similar latitudes (see Figure 4.8). In a few land areas, particularly those covered by lush tropical rainforests, evaporation plus the transpiration of moisture through plants can exceed evaporation over adjacent oceans, since the tree leaves present more total evaporating surface to the air than the ocean surface does. These are the exceptions rather than the rule, however. Such forested areas exist in equatorial regions with great amounts of cloudiness and high atmospheric humidity, which tend to reduce vapor pressure gradients, and hence evaporation. Among the ocean areas, the subtropical oceans overlain by the subtropical atmospheric high pressure cells experience the greatest evaporation, since the air there is generally clear and dry. Intense heating from the sun coupled with continual replenishment of fresh, dry air through sub-

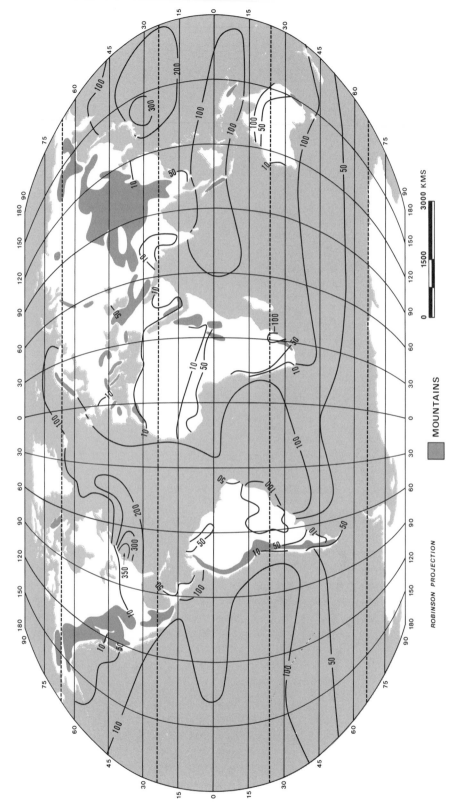

ROBINSON PROJECTION

MOUNTAINS

Figure 7.3. Evaporation in mm., January. (From Budyko, Atlas of Heat Balance of the Earth.)

Figure 7.4. Evaporation in mm., July. (From Budyko, Atlas of Heat Balance of the Earth.)

Figure 7.5. Mean latitudinal distributions of atmospheric vapor pressure and relative humidity. (From Howard J. Critchfield, GENERAL CLIMATOLOGY, *4th ed., © 1983, p. 45. Reprinted by permission of Prentice-Hall, Inc., Englewood Cliffs, N.J.)*

sidence from above produce low vapor pressure in the surface air relative to the underlying sea surface.

Nevertheless, the largest amount of evaporation on earth occurs at a somewhat higher latitude, off the east coast of North America between North Carolina and New Jersey. A similar situation occurs off the southeast coast of Japan, but it is not quite as pronounced. The high annual values in these areas are accounted for entirely by winter conditions. As can be seen in Figure 7.3, these areas are very conspicuous in January, but in July they do not appear at all (Figure 7.4). During winter, the vapor pressure difference between the water surface and the air above is maintained high, because the warm ocean currents flowing north-northeastward along these coasts are in stark contrast to fresh, cold, polar air that much of the time blows off the continents across the water currents. Since the Gulf Stream has a much closer contact with the mainland coast of North America than the Japan Current does with the mainland of Asia, the contrast is greater in the North American area, and hence the evaporation is greater. During summer when the air has about the same temperature as the water, or is even warmer, the evaporation is not outstanding in these areas.

Since the water vapor, once in the air, moves with the wind, its flux can be deduced crudely from annual wind direction and speed patterns relative to the major evaporation areas. Unlike the heat flux of the atmosphere, which moves simply from low latitudes to high latitudes with maximum values in the middle latitudes, the moisture flux has a more complex pattern, because the greatest amounts of water vapor are gained not at the equator but in the subtropics on either side. From these areas moisture moves both equatorward with the trade winds and poleward with the westerlies, reaching maximum flux around 10° and 40° latitude in each hemisphere, similar to the flux of latent heat shown in Figure 4.11.

The major part of the moisture flux in the atmosphere converges on the equatorial zone, so that on the average the specific humidity is highest at the equator and drops off steadily toward the poles, where the air is too cold to hold much moisture. The latitudinal distribution of relative humidity is quite different, however, since it depends not only on moisture content, but also on moisture-holding capacity. The lesser amount of specific humidity at high latitudes is still sufficient to bring the air near saturation in cool temperatures that reduce the capacity there to hold moisture. Therefore, average relative humidity is high both in the equatorial region and in the polar regions (Figure 7.5). It is lowest precisely in the subtropical regions of greatest evaporation, since moisture is moved outward in both directions from these zones by the winds, and temperatures are generally high.

Figure 7.6 shows the world distribution of average relative humidity in July. Here it can be seen that the subtropical high pressure zones stand out as belts of low relative humidity, while the equatorial zone between them, as well as parts of the middle and higher latitudes, mainly over the oceans, show higher relative humidities. The monsoon region of southeast Asia shows par-

ROBINSON PROJECTION

| <50% | 50–70% | 70–80% | 80–85% | >85% |

Figure 7.6. Mean relative humidity, July, in percentages. (After Száva-Kováts.)

0   1500   3000 KMS

ROBINSON PROJECTION

Figure 7.7. Mean precipitable water in mm., February. (After S.E. Tuller, "World Distribution of Mean Monthly Precipitable Water," Monthly Weather Review 96, no. 11 [1968].)

0   1500   3000 KMS

ROBINSON PROJECTION

0   1500   3000 KMS

Figure 7.8. Mean precipitable water in mm., August. (After Tuller, "World Distribution of Precipitable Water.")

ticularly high relative humidities at this time of year. In general, the interiors of middle-latitude landmasses, particularly in western North America and central Asia, are occupied by air with low relative humidities comparable to those in the subtropical deserts of the Sahara and Australia.

Maps of precipitable water (Figures 7.7 and 7.8) indicate the actual amount of moisture in the air in terms of amounts of precipitation that would fall if all the moisture in the air were precipitated. Here it can be seen that the intertropical convergence zone, which shifts latitudinally with seasons, generally has greatest amounts of atmospheric moisture, and amounts decrease poleward, with lowest amounts in the winter hemisphere. Areas with consistently converging air streams crossing warm ocean surfaces show particularly high values, such as northeast India during August.

The consequences of these various distributions of atmospheric humidity will become evident in subsequent chapters when atmospheric stability, storm development, clouds, and precipitation are discussed.

8 | ADIABATIC PROCESSES AND ATMOSPHERIC STABILITY

It has already been stated that temperature normally decreases with height at a fairly consistent rate throughout the troposphere. On the average this amounts to about 6.5°C per kilometer (3.5°F per thousand feet). One might question how such a situation can occur, with warm air below and cold air above. Since warm air is lighter than cold air it would seem logical that the warm air would tend to rise and the cold air to sink, thereby bringing about a stable situation that would resist motion. The answer lies in the vertical decrease in atmospheric pressure with height and the increase in potential energy of position above the earth's surface.

As the warm air rises, it comes under constantly decreasing atmospheric pressure, which allows it to expand. The work expended by the rising air to expand its volume and increase its potential energy above the earth's surface requires energy, which must be derived from some source. This comes from the internal energy (molecular activity) of the air itself, which reduces the molecular motion of the rising air, and hence, by definition, its temperature. This change of temperature without addition or subtraction of heat energy from external sources is known as an *adiabatic* process.

It can be demonstrated mathematically that the temperature of vertically moving air changes at the constant rate of ap- proximately 10°C per kilometer (5.5°F per thousand feet). This is assuming the air is not saturated and no latent heat is being released by the process of condensation. This is known as the *dry adiabatic lapse rate.* Since this rate is greater than the average, or environmental, lapse rate mentioned before, it can be seen that under normal conditions the temperature of a rising parcel of air will decrease more rapidly than the temperature of the air around it. So the rising parcel everywhere is colder than its environment, hence more dense, and tends to settle back down to the level from which it came. Thus, the average lapse rate is a stable condition in spite of the fact that the temperature decreases with altitude.

Temperature changes in the free atmosphere over short periods of time are essentially adiabatic. Since air is a poor conductor of heat, each air parcel tends to retain its own thermal identity. Therefore any parcel of air that is moving vertically can be expected to change its temperature at the adiabatic rate, decreasing 10°C per kilometer as it rises, or increasing 10°C per kilometer as it descends.

Over the long run this is not so true, and some heat energy is exchanged from one parcel of air to another. Thus the average condition is not adiabatic. Normally, temperature decreases upward through the atmosphere at a rate of about

6.5°C per kilometer, and the atmosphere will probably not be experiencing much vertical motion. This is the lapse rate that is used to change recorded temperatures to equivalent sea level temperatures when constructing temperature charts for the world. The differences in elevation of land surfaces are reconciled by adding 6.5°C for every kilometer (3.5°F for every 1000 feet) above sea level. Thus, if a station at 2 kilometers elevation records a temperature of 10°C, the temperature would be recorded on a sea-level equivalent map as

$$10° + (6.5 \times 2) = 23°C.$$

However, if an air parcel with an initial temperature of 10°C should descend 2 kilometers to sea level, its resultant temperature would be

$$10° + (10 \times 2) = 30°C.$$

Thus, atmospheric subsidence greatly warms the air and typically creates a temperature inversion (increase of temperature with height) at the bottom of the subsiding layer.

Near the surface of the earth, temperature changes in the atmosphere are primarily nonadiabatic, or diabatic, because of the intensification of other transfer processes, such as radiation exchanges, conduction, and turbulence. Superadiabatic lapse rates are common in air immediately above the earth's surface when the insolation is intense and the surface is dry, since sensible heat is added to the surface air more rapidly than it can be mixed upward. Such conditions do not usually extend very far upward into the atmosphere, although over very hot desert surfaces they commonly extend hundreds or even several thousands of meters. Such conditions typically produce extreme turbulence in the lower troposphere.

Rising air will eventually cool to its dew point, after which condensation begins. The level at which this takes place is known as the *condensation level* in the atmosphere

and is usually marked by the bases of cloud formations. Above this level latent heat of condensation will be released that will slow down the cooling caused by the lift and expansion of the air. Then the air cools at the *wet* (pseudo) *adiabatic lapse rate.* This is not a constant rate, since the amount of moisture condensed, and hence latent heat released, depends on the temperature of the air at which the process takes place. In the lower troposphere there is usually enough water vapor in the air to release a significant amount of latent heat when condensation is caused by rising air. But higher in the troposphere, the amount of water vapor in the air is so small that the latent heat release becomes insignificant even though the air is still saturated. Thus, in very warm air near sea level, the wet adiabatic lapse rate may be as little as 4°C per kilometer (2.2°F per 1000 feet), but in the middle troposphere the wet adiabatic lapse rate commonly approaches the dry adiabatic lapse rate. At a temperature of -40°C (-40°F), the wet adiabatic lapse rate is more than 9°C per kilometer (5°F per 1000 feet).

Atmospheric Stability

The stability characteristics of the atmosphere are very important, since they determine what kinds of vertical motions are likely to take place, and hence whether or not severe storms might develop. Absolutely stable air resists all vertical motion. Although it might be forced to rise by some external process, such as the wind blowing over a mountain range, it would never rise under its own volition, initiated only by energy contained within the air itself. Unstable air, on the other hand, would not resist motion, and once set in motion would continue that motion until changed by some outside force. Usually the state of the atmosphere is somewhere between these extremes. It might be potentially unstable, provided some outside force carried it through some process to a stage that crossed the threshold of instability within the atmosphere.

The stability state of the atmosphere is determined by the availability of energy sources that are functions of the temperature distribution with height and the water vapor content of the air, which represents an energy source in the form of latent heat. These two parameters (temperature and humidity) are observed through a vertical cross-section of much of the troposphere by a device known as a radiosonde—a small box filled with instruments to measure pressure, temperature, and humidity and a small radio transmitter to radio the information back to the ground as the box is carried aloft by a large hydrogen-filled balloon. The radio signals are received at the surface and translated into temperature and mixing ratio data for corresponding pressure levels. The upper winds may be measured at the same time by receiving the signals with directional antennas. The balloon normally rises to 15–20 kilometers, at which height it has expanded under reduced atmospheric pressure to its elastic limit, whereupon it bursts and falls back to earth. The radiosonde is protected from destruction by a small parachute that opens on descent.

The data thus gathered are used to construct (a) vertical profiles of temperature and humidity through the troposphere above given points on the earth's surface and (b) constant pressure charts that show heights, winds, and temperatures over large portions of the earth at standard pressure levels, such as the maps shown in Figures 6.13 to 6.17. They are also used to construct charts that illustrate the thickness of atmospheric layers between given pressure levels, which can indicate something about thermal winds and, over time, reveal areas of divergence and convergence, vertical motions, and changes in stability characteristics.

The vertical cross-sections of temperature and humidity above given points reveal the stability characteristics of the atmosphere. The degree of stability can best be illustrated by running through an example of the procedure used every day by weather forecasters utilizing some form of thermodynamic diagram. The diagram used here is the so-called "pseudo-adiabatic chart," which is composed of two principal coordinates, atmospheric pressure, represented by horizontal lines, and temperature, represented by vertical lines. Upon these two coordinates are superimposed three or four other sets of lines that are mathematically related to pressure and temperature (Figure 8.1).

The slanting light gray lines are known as the dry adiabatic lines. These slope at an angle that equals the dry adiabatic lapse rate, 10°C per kilometer (5.5°F per thousand feet). The slightly sloping darker-gray lines are the saturation mixing ratio lines. They show how many grams of moisture per kilogram of dry air are needed to saturate the air at any pressure and temperature shown on the graph. The dashed gray lines are the wet (pseudo) adiabatic lines representing the rate of change of temperature with height for saturated air that is moving upward and condensing moisture as it cools. The temperature then decreases less rapidly than the dry adiabatic rate, because the latent heat of condensation is being released into the air. At lower elevations (higher temperatures) the wet adiabatic lapse rate is much less than the dry adiabatic lapse rate (on the order of 4°C per kilometer) because of the great amount of water vapor being condensed. But at higher elevations (cooler temperatures) where saturation mixing ratios are low, the wet adiabatic lapse rate approaches the dry adiabatic lapse rate, as the latent heat of condensation becomes negligible. Thus, the wet adiabatic lines are not straight lines, but curve upward to the left on the graph. The sixth set of lines,

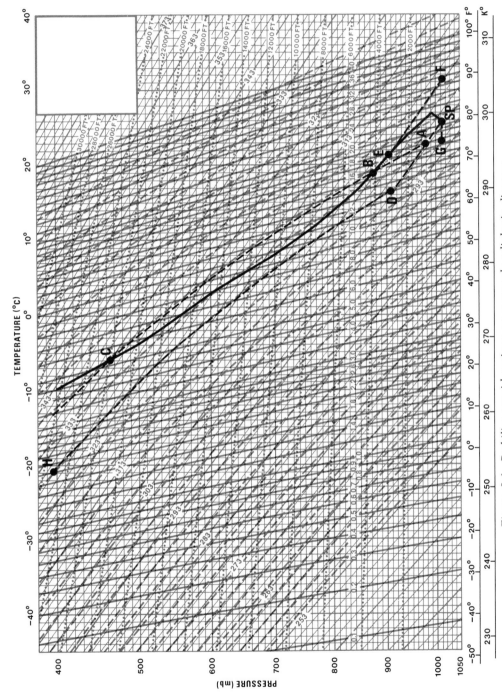

Figure 8.1. Stability considerations on a pseudo-adiabatic diagram.

the dotted gray lines that slope upward slightly toward the left, give a crude indication of the height above sea level according to standard pressure and temperature relationships.

The solid black line on the graph represents the actual lapse rate curve, at a given time over a given point, that has resulted from observations of pressures and temperatures at different levels by a radiosonde run. It can be seen that the surface temperature was 26°C (approximately 78°F) and there was a slight temperature inversion at the surface that extended upward for only a few hundred meters. Above the inversion for a short distance, the air had almost a dry adiabatic lapse rate, but then the lapse rate decreased and remained fairly consistent to the 400-millibar level, where the temperature fell to −10°C.

The moisture content was also recorded at each point, but for this first approximation of stability characteristics, only the mixing ratio at the surface point is needed. Let us assume initially that the actual mixing ratio measured at the surface point was 18 grams of moisture per kilogram of dry air. From this diagram relative humidities and dew points can easily be computed, because the equations that relate them to pressures and temperatures have already been represented graphically. It can be seen that the saturation mixing ratio for the surface point is approximately 21 grams per kilogram. Since it is known that the actual mixing ratio was 18, one simply divides 18 by 21 to get the relative humidity, which is approximately 86 percent. To get the dew point, one moves the surface point to the left at the constant pressure of 1010 millibars until the 18-gram saturation mixing ratio line is crossed, and then reads directly below, point G on the adiabatic diagram. It can be seen that the dew point is about 23.4°C, or almost 74°F. Since all three temperature scales are given on the graph, it is easy to convert from one temperature to another.

To determine the stability characteristics of the air under the process of mechanical lifting, raise the surface point through the rest of the air and observe how its temperature changes. Initially, as long as the air is not saturated, the temperature change parallels the dry adiabatic lines. But at point A on the graph, the temperature path of the rising surface parcel crosses the 18-gram line and the rising air becomes saturated. Above this point the temperature decreases at the wet adiabatic rate, as shown by the dashed line ABC. At point B the temperature path of the rising parcel crosses the originally plotted lapse rate curve. This means that the temperature of the rising parcel becomes warmer than the surrounding air, and therefore no more outside force is necessary to keep the surface parcel moving upward: above point B, which is known as the level of free convection, the rising air is buoyed up by the fact that it is warmer, hence lighter, than its environment at every level. Hence, the air no longer resists upward motion and becomes unstable. The motivating energy is the latent heat of condensation that is constantly being released as the air rises. The air, thus, is conditionally unstable. On the condition that some mechanical force (mountain range, front) is provided to lift the air to point B, the air becomes unstable, and a thunderstorm will likely develop.

The temperature path of the rising parcel recrosses the plotted lapse rate curve at point C, and once again the rising parcel becomes colder than its surroundings and resists upward motion. Assuming no outside force is available at this height, upward motion will cease at point C, and this will mark the top of cumulus-type clouds. The temperature path of the rising parcel recrosses the plotted lapse rate curve because the wet adiabatic lines are curved upward to the left. If the capacity of the air to hold moisture decreased at a constant rate with decreasing temperature, such would not happen, and theoretically there would be no limit to the upward motion of the surface air, hence no upper limit to the buildup of clouds. But since the capacity of the air to hold moisture decreases at a decreasing rate with decreasing temperature, the amount of moisture, and hence the latent heat release, eventually becomes negligible with cold temperatures, and an upper limit to instability is thus set.

To indicate the significance of moisture content to stability, let us assume that the surface air has a mixing ratio of only 13 grams per kilogram. As the surface air parcel rises the temperature will again initially change at the dry adiabatic rate. But this time the temperature path of the rising parcel will parallel the dry adiabatic lines up to point D, where the 13-gram line crosses. The condensation level, and hence the base of the cloud formation, has now lifted from A to D. Above D the air will condense moisture and the temperature will change at the wet adiabatic rate, line DH. The temperature path of the rising parcel never crosses the original plotted lapse rate curve. Hence, the air resists motion all the way up and never becomes unstable; the air is now absolutely stable. It could still be forced to rise by some outside mechanism, such as a mountain range or a front, when clouds would form above the condensation level, and perhaps even precipitation would take place; but the air will always remain stable, and the cloud type now will be stratus rather than cumulus.

The previous two examples assume that a mechanical lift initiates the upward motion of the surface air. Nevertheless, strong heating at the surface may cause convective activity to extend from the surface upward to the level of free convection, above which the air continues on an unstable course. To test for that, mark the point where the mixing ratio of the surface air crosses the plotted lapse rate curve. In our initial example, the 18-gram line crosses the plotted curve at point E. This is the condensation level for the surface air if lifted by con-

vective currents only. It also now becomes the level of free convection, since above that point the air will follow a temperature path to the right of the plotted lapse rate curve. To see how high the surface temperature would have to get for the air to be lifted to point E, drop a line parallel to the dry adiabatic line to the surface pressure at point F. An adiabatic lapse rate has now been hypothetically created from F to E along which any heated air can move without resistance, because it will always be at the same temperature as its surroundings at each level in its rising path. At point E condensation begins, the latent heat of condensation is added to the air, and the temperature path of the rising parcel then parallels the wet adiabatic curves, which puts it to the right of all the other curves on the graph (not shown).

Thus, this air could be made unstable if the surface temperature were raised to about 32°C (89°F). With lift due to surface heating, the condensation level is higher than it was with mechanical lift, because the air is first heated and then cooled adiabatically as convective currents set in. But the condensation level now is also the level of free convection, which is lower in the atmosphere than it was with mechanical lift. Therefore, instability begins at a lower level and extends to a higher level, since the temperature path of the rising parcel is now farther to the right on the diagram. This is the kind of instability that may occur on a hot summer day over southeastern United States in warm, moist air coming from the Gulf of Mexico. Typically, this kind of air produces many scattered thunderstorms throughout humid summer afternoons.

Initially stable air may be made unstable by lifting a thick layer that has a large humidity lapse rate. The wet air in the lower part of the layer may reach its condensation level before the drier upper part of the layer does, which means that with continued lift the lower air cools at the wet adiabatic rate, while the upper air is still cooling at the dry adiabatic rate. In essence, the lower air is being warmed relative to the upper air by the release of the latent heat of condensation. Such a condition is known as *convective instability* and can be responsible for severe thunderstorm turbulence and tornado formation in areas such as the Great Plains of North America where, during summer, dry air from the southwestern deserts frequently overruns moist air from the Gulf of Mexico.

Lifting of a thick layer of air will decrease stability even if the moisture content does not vary throughout the layer. Since the atmosphere becomes less dense with height, the top of a column of air must rise more than the bottom to experience the same pressure decrease. Since adiabatic temperature changes are directly related to changes in height, not pressure, the top of the column will cool more than the bottom. Thus the lapse rate will be steepened. The opposite is true with descending (subsiding) air. Active subsidence typically produces a temperature inversion, hence extreme stability, at the base of the subsiding column.

9 | AIR MASSES AND FRONTS

Air Masses

Since the air in the lower troposphere outside the equatorial doldrums circulates primarily around cells quasi-parallel to isobars, at least above the friction layer, such cells become persistent through time, and the air within them becomes thoroughly mixed and homogeneous in the horizontal plane and distinct from adjacent circulations with which there is little air exchange. This is particularly true in high pressure cells, where friction at the surface causes the air to diverge gradually away from the center of the cell, thus inducing subsidence in the vertical dimension and a spreading out of similar air over a large area at the surface. On the other hand, low pressure cells at the surface are areas of convergence of different kinds of air from various points on the periphery of the circulation system, and therefore usually do not lead to homogeneity of air over a wide area.

Some cells of circulation are persistent in their configuration and location. Some are even permanent, such as the subtropical high pressure cells over the oceans of both hemispheres. Others are more transitory, retaining their characteristics only for a number of days, or weeks at most; and during that time they may move long distances and come into contact with varying surfaces. Such shifting of positions brings about exchanges of heat and moisture between the surface and the air, which ultimately modify the air significantly.

Air within a circulation cell that shows a relatively high degree of homogeneity over a broad horizontal area is known as an air mass. It is distinguishable by its temperature, moisture, and stability characteristics, which, in combination with its temperature relative to the earth's surface, produce predictable associations of weather. Of course, if a circulation system occupies an area hundreds, or even thousands, of kilometers in radius, the air will not be completely homogeneous from one side to another. Temperatures will be higher on the equatorial edge of the air mass than on the poleward edge, and moisture and stability characteristics may vary also. Nonetheless, there is much less variation within an air mass than between it and adjacent air masses.

Over very broad circulation systems, such as the subtropical highs over the oceans, very different types of air may develop at opposite ends. For example, the air on the eastern end of the North Atlantic subtropical high has circulated for thousands of kilometers across the middle and northern portions of the North Atlantic over relatively cool water and has then turned southward around the eastern end of the high, where it has undergone subsidence that has produced a strong temperature inversion at or near the surface. Such conditions produce extreme atmospheric stability, rainlessness, cool surface temperatures, near atmospheric saturation, and much fog and low stratus clouds. As the air turns westward it circulates across thousands of kilometers of very warm water, which injects heat and moisture into the air and lifts the inversion layer on its westward course. As it turns northward around the western end of the high, the air is very warm with a high mixing ratio and unstable or conditionally unstable characteristics that induce much vertical turbulence, cumulus cloud buildup, and thundershowers. The surface air is warm and muggy, with high temperatures combined with high relative humidities. Since the subtropical highs tilt eastward and equatorward, the strongest subsidence is on the eastern-equatorward periphery of the cell, and the weakest subsidence or uplift is on the western-poleward side of the cell.

Source Regions

Two conditions are necessary for air masses to form: (a) a large, relatively homogeneous portion of the earth's surface, and (b) air stagnation or long trajectory over that surface sufficient for a significant thickness of the air to take on the characteristics of the surface. Exchanges of heat and moisture must take place between the earth's surface and the air above so equilibrium between the two can be reached. This takes time—at least several days, and perhaps weeks. Portions of the earth's surface that satisfy these two conditions are known as air mass source regions. Examples are the oceanic areas dominated by the subtropical high pressure cells and certain land areas. Particularly in winter, when they do not receive much sunlight, large portions of the land in high latitudes are covered with snow that presents a fairly homogeneous albedo and radiation emission. Certain desert areas may act as source regions during summer months.

Air may tend to stagnate over certain areas of the earth's surface because of the character of the general circulation of the atmosphere, which forms nodes of circulation in some latitudes and in some land-sea and topographic configurations. Stagnation may take place in other areas because of interior locations between mountain ranges, and so forth. Examples are the intermontane region in southwestern United States and northern Mexico which develops very hot dry air during summer, and the similar Central Asian area east of the Caspian Sea.

During winter the Mackenzie River valley in northwestern Canada receives streams of air from various directions off the Arctic and Pacific oceans, after which the air stagnates and cools over the snow-covered land surface. This process ultimately results in a cold, stable air mass with a low mixing ratio that may extend two or three kilometers into the air and produce high surface pressures. When the air pressure has built up enough, the air mass typically bulges southward, and a cell buds off from the main air mass to move out of the source region southeastward into central United States. Such a discrete cell may separate from the main body of an air mass and move en masse thousands of kilometers downstream, where it slowly modifies over continuously changing surfaces and eventually loses its identity.

Sometimes broad streams of air flood areas downstream with a given type of air without forming discrete circulation cells apart from the main bodies of parent air masses in their source regions. Such are the warm, moist, unstable air masses that move poleward along the western edges of subtropical highs and the relatively cool, stable air masses that move equatorward along their eastern edges.

Air masses are designated according to their source regions as being maritime or continental in origin, and equatorial, tropical, polar, or arctic (or antarctic). Thus, an air mass may be designated as mT (maritime tropical), cT (continental tropical), mP (maritime polar), cP (continental polar), and so forth. In the middle latitudes where the contrasts between adjacent air masses are greatest, the primary air masses are mT, cP, and mP, with cT significant in dry land areas. Such designations have become rather standard, although some of them are sort of misnomers. For instance, the polar air masses do not originate at the poles, but in the subpolar and higher middle latitudes, and are typically found throughout the middle latitudes, while the polar areas are dominated by arctic or antarctic air. Soviet climatologists refer to temperate air masses in the middle latitudes rather than to polar air masses, but this is an ambiguous designation also, since the word "temperate" has been used to encompass a multitude of conditions.

Weather Associations

Air masses are typically associated with certain types of weather that correspond to their water content, stability characteristics, and temperatures relative to the surfaces over which they are traveling. For instance, cP air during midwinter in its source region will have taken on the cold temperatures of the underlying snow-covered surface, which will have led to low mixing ratio content, the formation of a temperature inversion based at the surface, and extreme atmospheric stability. In spite of the fact that the surface temperature inversion with its resultant stability essentially rules out convectional mixing, the cooling effects from the surface continue to extend upward into the air, because of the peculiar reflective and radiative properties of snow. Fresh snow reflects at least 90 percent of incident short-wave radiation, which is already minimal at this time of year at high latitudes. And it radiates almost perfectly (as a black body) at longer wavelengths in the infrared, where all the terrestrial radiation is because of cold temperatures. Thus, the snow assures that little heat is received and much heat is lost.

There is net radiation upward until very low surface temperatures are reached (commonly $-50°$ – $-70°C$), despite the fact that the air one or two kilometers above the surface is commonly 20° to 30°C warmer than the snow surface. The snow radiates heat upward, and the warm inversion layer of air radiates heat downward into the colder air below. Because of its colder temperature, the snow radiates heat less intently at any wavelength than does the warm air above. But the snow radiates heat across a full spectrum of wavelengths according to its temperature, while the atmosphere, being made up of a mixture of gases, radiates heat only at certain wavelengths that correspond to the radiation and absorption characteristics of the gases in the air. Therefore, the spectrum of radiation coming from the upper air is far from complete, and the sum total of radiation upward from the snow surface is greater until the temperature of the snow surface reaches very low values. After that, an equilibrium is reached, and surface air temperatures remain essentially constant until some advection of different kinds of air upsets the situation, when the process is repeated to reestablish an equilibrium temperature.

Such equilibrium temperatures are generally reached initially in December in places such as northwestern Canada and Siberia, and from then on the temperature varies little until it begins to warm up again with increased insolation as the spring equinox is approached. Thus, the inhabitants of Siberia talk about their "coreless" winters, which do not have temperatures arranged symmetrically about a minimum point in January or February but, instead, consistent temperatures with some nonperiodic variations from December, or even late November, through early March.

When cP air moves out of its source region in northwestern Canada southeastward into the United States it moves across a surface with (usually) constantly increasing temperatures that are everywhere warmer than the surface temperature of the air. Thus, it is classified as a "k" (cold) air mass. The term "cold" is relative, meaning simply that the air mass is colder than the surface over which it is traveling. The air mass will be modified as it moves over the warmer surface, receiving heat, and probably moisture, from the underlying surface as it goes. The heating at the surface will cause convective currents to set up in the lower air, which cause a vigorous mixing of moisture and pollutants upward, making good surface visibility.

If the air is much colder than the surface, the convection may be vigorous enough to produce small cumulus clouds with well defined bases at a common level—the condensation level of the surface air—and tops that are limited by the high degree of stability of the upper air, which has not yet been modified from its cP origin. Generally such cumulus are no more than a few hundred meters thick, with much blue sky in between individual clouds. They have become known as "cumulus of fair weather," although this may seem a contradiction in terms, since if there are clouds the weather is not entirely fair. The connotation is that these cumulus are not going to result in much because the air a few hundred meters above the ground is very stable. They are produced entirely by daytime heating at the surface and will dissipate soon after sundown. Such is the weather usually associated with k air masses—crisp, clear surface visibility, and cumulus of fair weather.

Occasionally, if the temperature differential between a k air mass and the surface over which it is traveling is very great, and if enough moisture is available, the cumulus buildup may reach showery proportions, and the weather for the day will be marked by sharp bursts of showers that last only a few minutes at a time, interspersed by clear, sunny weather in between. One short shower after another may continue through much of the day. This is particularly common in the U.S. Midwest during spring,

when the cP air coming south from Canada is still cold and the ground surface over which it is moving has already warmed up greatly. If the temperature is cold enough there may be snow showers, but this is more common in the Midwest during the onset of winter, when the first cold blasts of Canadian air move over a still relatively warm surface. Graupel (snow pellets) are common occurrences with such weather. The showery weather, whether it is rain or snow, is known as instability showers, since it is caused by low-level instability due to surface heating. Such weather is commonly associated with the leading edge of a continental polar outbreak in midwestern United States after the passage of a cold front associated with a cyclonic storm.

Figure 9.1 shows a typical lapse rate curve for a cPk air mass moving southward out of Canada into midwestern United States. Initially the entire air mass is quite stable, as illustrated by line D-C and above. but as the air moves southward over warmer land the surface air heats to a temperature of 10°C, as shown at point A, which establishes a dry adiabatic lapse rate upward to about 850 millibars (1.6 kilometers above the surface). The surface air can now move freely throughout this lowest layer, with moving air parcels remaining at the same temperature as their surroundings at each level. At point B the surface air, with a mixing ratio of 6 grams per kilogram, becomes saturated, and above that point it cools at the wet adiabatic rate. This causes an unstable layer between points B and C, above which the rising air once again becomes colder than its surroundings. Thus, cumulus of fair weather are formed with their bases around 1500 feet (point B) and their tops around 7500 feet (point C). They cannot grow any taller because the upper air is still quite stable, not having been modified by processes going on at the surface. Often a small letter "s" is added to the symbol cPk to designate that the air is still stable aloft (cPks).

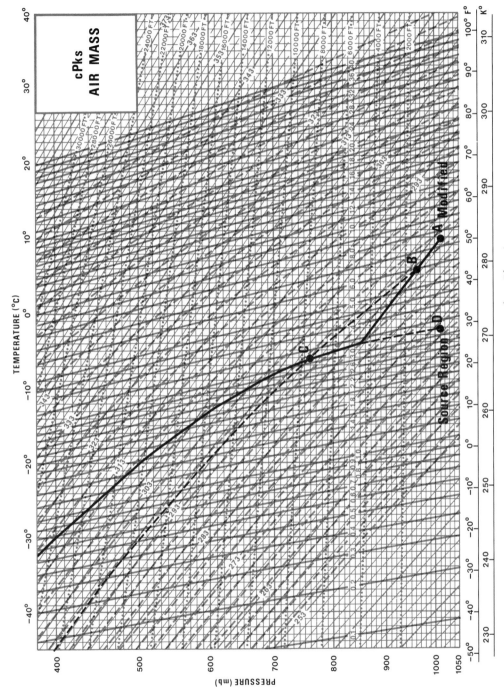

Figure 9.1. Vertical structure of a cPks air mass.

Maritime tropical air moving up the Mississippi Valley from the Gulf of Mexico is usually a "w" (warm) air mass, particularly during cooler seasons. Modification takes place by heat loss from the warmer air to the cooler surface, thereby stabilizing and stratifying the surface air. This often leads to poor surface visibility, haze, and even fog, or large amounts of dew, depending upon the motion of the air. The mT air is commonly warm, humid, and perhaps conditionally unstable, but the cooling at the surface stabilizes the surface air and rules out convective activity.

During summer, maritime tropical air moving northward up the Mississippi Valley from the Gulf of Mexico may actually be a "k" air mass, even though its temperatures may be more than 30°C (86°F). The air over the Gulf at that time of year typically has surface temperatures of 86°F but the land in the Mississippi Valley, perhaps even as far north as Wisconsin and Minnesota, may be between 90° and 100°F, or even above 100°. Thus, as the air moves northward from the Gulf it is heated at the surface; this process may easily set off convectional thunderstorms since the air commonly is conditionally unstable and has such high relative humidity that it does not have to rise very high to reach the condensation level, after which the latent heat released is usually enough to buoy the air on upward.

Figure 9.2 illustrates a typical maritime tropical air mass moving northward from the Gulf of Mexico up the Mississippi Valley. Initially, over the Gulf, its surface temperature is 30°C (86°F), as shown at point A. In late spring or early summer as it moves northward over the land it encounters cooler temperatures, and the surface air temperature cools to 26°C, illustrated by point B, which is below the dew point, assuming a surface mixing ratio of 22 grams per kilogram. This produces a surface temperature inversion, surface stability, and poor visibility, perhaps fog.

Later in summer, the same air mass might prove to be an mTk air mass as it encounters warmer temperatures over the land as it moves northward. The surface temperature might warm to 37°C (98°F), which would then produce a dry adiabatic lapse rate from point C up to point D, the condensation level where the 22 grams of moisture in the air saturates the air. Point D also becomes the level of free convection, since the rising air (dashed line) above that point everywhere is warmer than its surroundings. Since the initial lapse rate curve slopes at an angle intermediate between the wet adiabatic lines and the dry adiabatic lines, the air mass is conditionally unstable, and surface heating during a hot summer afternoon may be enough to trigger thunderstorms. If the surface temperature reaches 37°C or above, this will happen.

In the first instance, the cooler land stabilized the lower 2000 feet of the air mass, but the air still remained conditionally unstable aloft. This conditional instability was released in the second instance when the warmer land heated the surface air. Thus, the same air mass can be either a k air mass or a w air mass, depending upon its relation to the temperature at the surface, and very different types of weather will result. In either case, the small letter "u" might be added behind the mTk or the mTw, to indicate unstable conditions aloft.

Continental tropical air, such as that which forms in southwestern United States during summer and in extensive subtropical deserts such as the Sahara, has exceedingly high surface temperatures and low relative humidities. The great heating at the surface causes strong convectional activity up to perhaps two or three kilometers above the ground, but above that extensions of subtropical highs generally predominate and form temperature inversions that suppress convection. Since the surface air has such low relative humidity, the convective activity usually is not enough to raise the surface air to its condensation level, and therefore the air remains cloudless even

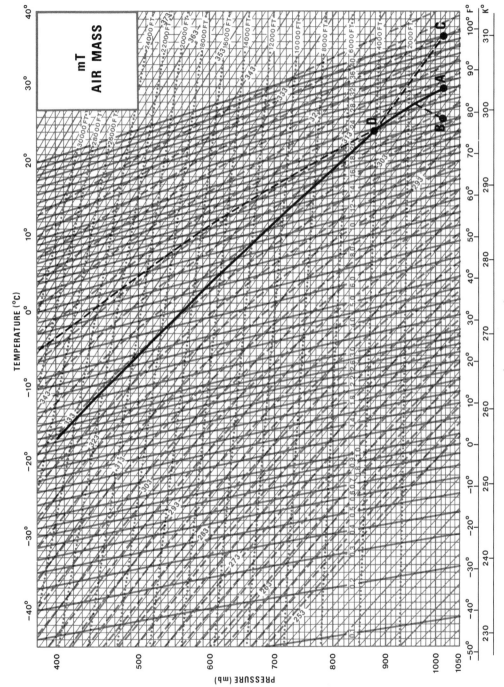

Figure 9.2. *Vertical structure of an mT air mass.*

though the first two or three kilometers of air are extremely turbulent. cT air is usually a summer phenomenon only.

Maritime polar air moving around the poleward edges of the subtropical high pressure cells is typically cool, moist, and stable as it moves equatorward along the western coasts of North America, Europe, and the Southern Hemisphere continents in the middle latitudes. If it penetrates far enough inland, as it does over Europe, it may modify into a fairly unstable air mass, with thundershower activity during summer, or become colder, drier, and more stable, similar to cP air during winter. mP air is usually a k air mass in summer and a w air mass in winter. Such air does not penetrate far inland on the North or South American continents, since in both cases high mountains rim the Pacific.

Maritime polar air dominates the west coasts of these continents much of the year and brings much stratus clouds and fog to the coastal areas, but thunder is seldom heard on the west coast of North America. This may change as the air rises up the windward slopes of mountains, such as the Cascades and the Sierra Nevada in Washington, Oregon, and California, where the air may eventually reach its level of free convection and thereby spawn thunderstorms, the lightning from which frequently sets off forest fires. But by the time the mP air has crossed such mountain ranges and descended their leeward slopes, it is so modified that it can no longer be identified as maritime polar air. The greatest stability of maritime polar air is found in the equatorward extremities of the eastern ends of the subtropical highs adjacent to such coasts as southern and Baja California, Peru, and northwestern and southwestern Africa.

Fronts

Although certain types of weather are associated with air masses, as described above, most weather (other than clear weather) is associated with boundary zones between air masses. Since most identifiable air masses are associated with high pressure cells, the typical weather within air masses, particularly in their source regions before they begin to be modified by changing temperatures at the surface, is clear, calm weather associated with subsiding air. Although scattered showers may be associated with k air masses or fog with w air masses, these are generally mild weather phenomena compared to the well-organized and intense cloud systems, turbulence, and heavy, prolonged precipitation that may be associated with boundary surfaces between air masses where warmer air is forced to rise over colder air. The colder air mass acts as a wedge over which the warmer air rises, expands, and cools adiabatically, which reduces its capacity to hold moisture to the point where condensation must take place, clouds form, and precipitation falls. Such activity generally produces broad areas of continuous overcast skies and precipitation, and if the warm air is conditionally unstable it may also produce extreme turbulence manifested by thunderstorm activity. Because most hazardous weather is found along these boundary zones rather than within the air masses themselves, weather forecasters generally concentrate most of their attention on the boundary zones, particularly in the middle latitudes where air mass contrasts are greatest and, hence, fronts are most intense.

These boundary zones have become known as fronts because their advances and retreats between air masses reminded the Scandinavian meteorologists of the day-to-day shifts in the battle fronts in western Europe during World War I, at the time when the air mass and frontal theory of the atmosphere was being developed. On a surface weather map fronts are shown simply as lines, but it must be remembered that they are zones, usually several tens of kilometers wide, and they are curved surfaces that extend upward into the atmosphere. The line on the surface weather map represents only the intersection of the frontal surface with the ground.

Distribution

There are three main frontal systems on earth. Their approximate mean positions for January and July are shown on Figures 9.3 and 9.4, along with the air mass types they separate. They shift seasonally, along with the wind and pressure belts, as a result of the seasonal shift of insolation across the face of the earth. And within seasons they shift drastically, so that any daily map might look quite different from these average maps.

1. The first of the three frontal zones is the *intertropical front*, or better termed the *intertropical convergence zone* (ITCZ), since it is not a very distinct front in terms of air mass contrasts across it. It is a broad zone of convergence between the northeast trade winds in the Northern Hemisphere and the southeast trade winds in the Southern Hemisphere. Both of these wind systems carry maritime tropical unstable air masses, so there is little contrast in temperature, humidity, and stability conditions across the front. But because both air masses are warm, moist, and unstable, they have greater capabilities for producing clouds and precipitation, and copious amounts of precipitation fall along this front more than any other. The ITCZ is generally in an area of little surface observational record, with much of the area over oceans, so not much is known about its structure. It is often a diffuse area of convergence with little discernible frontal surface. Probably much of the time the converging surface air is rising approximately vertically without any perceptible frontal slope. In these tropical regions of little surface weather information, the front is often located on daily weather maps simply by the associated cloud bands and precipitation, observed

from satellites which, of course, is the reverse of the intent for locating fronts on maps—to understand and predict the distributions of clouds and precipitation.

As can be seen in Figures 9.3 and 9.4, the convergence zone is generally continuous over the oceans and fairly constant in its location, particularly in the Pacific and Atlantic oceans. In the Indian Ocean its latitudinal position fluctuates widely by season, because of the unusual development of the monsoons in this part of the world. During summer the intertropical convergence zone is pushed north of the Tropic of Cancer in northern India, Indonesia, and southern China by the strong southwesterly monsoon winds that blow into thermal lows over the Asian continent at that time of year. The convergence zone is broken across equatorial South America by the high Andes. In both South America and south Africa during the Southern Hemisphere summer (January), the convergence zone tends to become oriented more north-south than east-west in the interiors of these land masses. Over the Pacific and Atlantic oceans, the ITCZ tends to lie north of the geographical equator at all times because of the generally stronger atmospheric circulation in the Southern Hemisphere.

2. The second set of fronts is the *polar front* in each hemisphere. This again is a misnomer, since these fronts are in the middle latitudes rather than in the polar regions. They are by far the most distinct fronts in terms of air mass contrasts. They generally separate cP air from mT air, particularly in the continental areas of the middle latitudes of the Northern Hemisphere. In oceanic areas they sometimes separate one variant of mP air from another variant of mP air. This may also occur over land areas where maritime polar air has free access, such as in western Europe, where the polar front typically separates fresh mP air on the west from a modified form of mP air on the east. The polar front may also separate some variant of mP air

Figure 9.3. Mean positions of surface air masses and fronts, January.

Figure 9.4. Mean positions of surface air masses and fronts, July.

from some variant of mT air. This usually occurs when the polar front moves into lower latitudes, almost always during winter when, of course, the mT air is not very classical tropical air.

The polar front is very discontinuous and distorted in its orientation. Its position fluctuates latitudinally much more than the intertropical convergence zone; at any season it may be found anywhere in North America from southern Canada southward across the entire United States into the Gulf of Mexico and Central America. The polar front's presence is much less felt in the interior of Asia. In the Southern Hemisphere it is more continuous around the earth, since there is little land in the middle latitudes. But the air mass contrasts across it are usually less than in the Northern Hemisphere, since all air masses associated with it are usually some form of maritime polar. Nevertheless, since all the air masses are of a maritime origin, the polar front in the Southern Hemisphere spawns much storminess, cloudiness, and precipitation.

3. The third set of fronts includes the *Arctic and Antarctic fronts* in the two hemispheres. They usually separate some variant of continental polar or arctic air from some variant of maritime polar or arctic air. They are very discontinuous and difficult to locate. Like the intertropical convergence zone, they frequently lie in uninhabited areas where little information on surface air pressures and winds is available. But unlike the ITCZ, the Arctic front does not produce much precipitation, since the air masses associated with it always have low temperatures and therefore little capacity for holding moisture. Segments of the Arctic front may spawn cyclonic storms that may produce rather frequent precipitation, usually in the form of snow, but never very much in amount. Segments of the Arctic and Antarctic fronts usually form somewhere near the coasts of northern North America, Eurasia, and the Antarctic. In those positions they lie in regions of con-

trasting surface temperatures that in summer are considerably warmer over the land than over the adjacent seas, and in winter are colder over the land than over the adjacent seas. Therefore these coastal areas are quite windy and may experience numerous blizzards.

Of all the fronts, the polar front has attracted by far the most attention, since it lies in the middle latitudes and strongly affects many of the more-developed countries of the world. The typical characteristics of fronts described in the following section relate specifically to the polar front, particularly in North America.

Characteristics

Fronts are labeled "cold" or "warm" according to the direction they are moving. If the cold air is advancing at the surface at the expense of the warm air, then the front is a cold front; if the warm air is advancing at the expense of the cold air, it is a warm front. If the front is stalled or, more likely, oscillating back and forth with little net change in position, it is called a stationary front. In any case, the warm air is being forced to rise over the cold air. The warm air is cooled adiabatically by the lift, and the resultant condensation forms, cloud formations, turbulence, and precipitation types and amounts relate primarily to the characteristics of the warm air mass. They also relate somewhat to the vertical structure of the front and the atmospheric circulation associated with it.

Although the type of front is determined by the direction of motion, each type of front tends to take on a certain form and associated weather phenomena as a result of the characteristics of the motion taking place. A cold front tends to have an initial steep slope that tapers rapidly aloft and usually reaches heights of no more than three kilometers, while a warm front tends to have a more-constant slope and eventually extends to considerably greater heights. The difference in the initial slopes of the two fronts may be due largely to frictional effects between the air and the ground. Since the cold air under a cold front is pushing forward, it is retarded by friction with the ground, while the cold air under a warm front is being dragged back, and friction with the ground tends to elongate its surface extent.

Slopes of fronts are usually no more than 1 or 2 percent, if that, which would be impossible to show in a small diagram drawn accurately to scale. Thus, in the next chapter, figures showing vertical cross sections of fronts will greatly exaggerate the vertical dimension with respect to the horizontal. Although the air rising up a frontal surface may move 100 kilometers or more horizontally while moving only one kilometer vertically, vertical motion is the significant one, since it produces adiabatic cooling of the warm air and resultant condensation and associated phenomena.

FLOW PERTURBATIONS IN THE MIDDLE LATITUDES

Waves

A frontal surface is a zone of relatively rapid change in atmospheric density. Although it is not an absolute discontinuity, such as the interface between the atmosphere and a standing water body, it nevertheless acts similarly in producing wave motion. Just as the wind whips up waves on the ocean surface, so also do differential air movements along frontal surfaces within the atmosphere. But since most of the air motion is horizontal, the waves created within the atmosphere along frontal surfaces tend to have much greater horizontal dimensions than vertical ones.

Figure 10.1 illustrates the life cycle of wave formation along a frontal surface in the Northern Hemisphere. Initially, a straight front separates cold air to the north from warm air to the south. In addition to the temperature contrasts and perhaps differences in other characteristics, such as humidity, the front separates two opposing air flows that tend to set up waves along the frontal surface. Around the apex of the wave, a counterclockwise flow develops that serves as a sort of third cog to take up the friction between the opposing air flows, as shown in portion C of Figure 10.1. The right half of the front now moves as a warm front, with the warm air advancing at the expense of the cold air, while the left half of the front moves as a cold front with the

cold air advancing. As the counterclockwise circulation intensifies, the wave increases in magnitude and shortens in length while the cold front catches up with the warm front (D).

In part E an occlusion process begins, with one front riding up the slope of the other front, which squeezes the warm air completely off the ground in the apex of the wave. The occlusion produces maximum uplift of the warm air and thereby creates the most intense stage of the storm; but the occluded stage is also the beginning of the end of the wave formation, since the air mass contrast across the front at the ground tends to be destroyed by the occlusion process. Cold air is now on both sides of the front at the surface.

Either side can be the colder, depending upon the situation. Often the western side is colder because that is the region of most active intrusion of fresh air from the cold air mass, the area where the cold air has had the least amount of time to be modified by contact with the ground. In certain instances, however, the air on the eastern side of the front may be colder because it is typically farther poleward in association with waves that are moving from southwest to northeast. If the air to the west under the cold front is colder than the air to the east under the warm front, the warm front rides up the slope of the cold front and the occlusion is a cold-front-type occlusion,

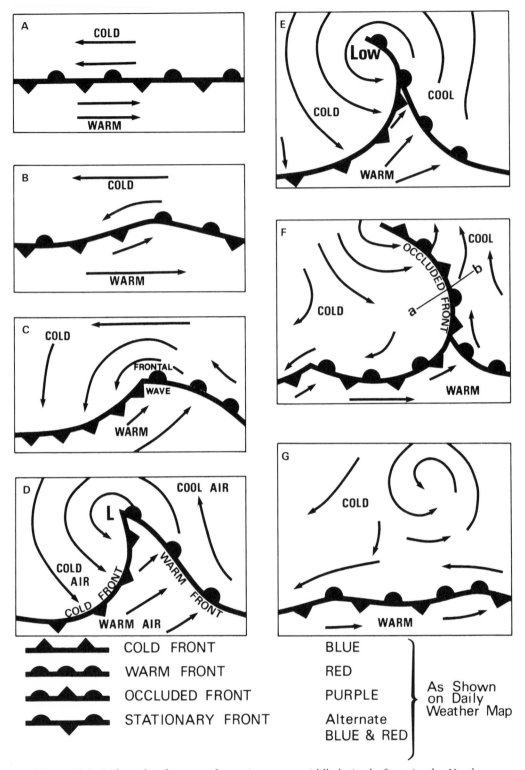

Figure 10.1. Life cycle of a wave formation on a middle-latitude front in the Northern Hemisphere.

since the cold front stays on the ground. Figure 10.2 shows this in a vertical cross section along the line ab in part F of Figure 10.1. But if the air on the eastern side of the occlusion under the warm front is colder than the air on the western side of the occlusion under the cold front, the warm front stays on the ground and the occlusion is a warm-front-type occlusion (see Figure 10.2).

As the warm air is lifted higher and higher by the occluding process, the occlusion begins to die (*frontolysis*) at the extreme tip and dissipates gradually toward the original wave apex (part G, Figure 10.1). For a time the counterclockwise circulation continues, and thin clouds remain aloft in the contact zone between the lower cold air and the warm air above it. But eventually all this dissipates and only the original front remains. By this time a secondary wave may appear along the trailing stationary front upstream from the previous wave. This often forms as the old occluded low pressure center begins to dissipate (part F). This wave will likely grow and occlude to repeat the cycle of the previous wave. In so doing it may move up the front to join the parent low pressure center. Frequently it traverses approximately the same territory, perhaps at a little lower latitude than the original wave.

Cyclones

When a low pressure center around the apex of a wave becomes encircled by closed isobars, it is known as a cyclone. This is primarily a phenomenon of the middle and higher latitudes, where air mass contrasts are great enough to produce distinct fronts and the Coriolis Force is strong enough to cause a rotary motion. In some instances the term "cyclone" has been used ambiguously. In interior North America pioneers built "cyclone cellars" (underground shelters from tornadoes) and confused the term "cyclone" with the term "tornado," which

refers to a violent, small whirl of wind. In the Bay of Bengal area east of India, the term "cyclone" has been used to denote a hurricane-type storm. Within the context of this book, unless otherwise specified, "cyclone" will refer to middle- and higher-latitude low pressure centers with frontal systems in them.

A typical mature, partially occluded Northern Hemisphere cyclonic storm is shown in Figure 10.3. The type of weather varies between different sectors of the storm, and the type of weather that can be expected as a cyclonic storm approaches depends upon the part of the cyclone that is going to pass through one's location. The most sustained lift of warm air occurs up the warm front, since the flow of warm air in the warm sector of the storm between the fronts is usually directly up the slope of the warm front. This produces a large area of continuous overcast skies and probably precipitation, usually extending 200–300 kilometers ahead of the surface position of the warm front. Thus, the northeastern quadrant of a cyclonic storm in the Northern Hemisphere is usually the cloudiest and the one that will produce the most prolonged and widespread precipitation.

Surface winds in this portion of the storm are from the east or southeast, a well-known sign of an approaching storm in the middle latitudes of the Northern Hemisphere and one that has often befuddled laymen into thinking that the worst weather in these regions comes from the east. These easterly winds are a part of the storm itself, however, and do not denote the direction of movement of the entire storm. One must keep in mind that there are two motions to be considered, a circulatory motion around the low pressure center, and a translational motion from some westerly direction to some easterly direction, typically from southwest to northeast in the eastern portion of the United States. Thus, in Figure 10.3, although the first harbingers of the storm are

A.

B.

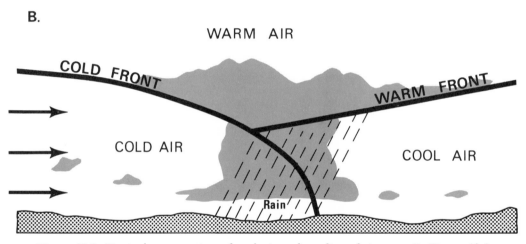

*Figure 10.2. Vertical cross-section of occlusion along line ab in part F, Figure 10.1.
(A) Warm-front-type occlusion; (B) cold-front-type occlusion.*

easterly surface winds, the storm is moving in from the southwest.

The southeastern portion of a cyclone is the warm sector. In eastern United States the warm air mass is some variant of maritime tropical air with relatively high temperatures, high humidities, and conditional instability. Weather in this sector is typified by scattered clouds, commonly of a cumulus nature, and associated scattered showers, perhaps thunderstorms. But

most severe thunderstorms are generally located along the cold front. Since this front has an abrupt lower slope, the warm air undergoes its most active rise here, which frequently triggers instability that produces severe thunderstorms, often in a continuous line all along the front. But since the front tapers off rather rapidly in the upper air, these storms do not extend far behind the front, typically no more than 30–40 kilometers at most, beyond which the weather

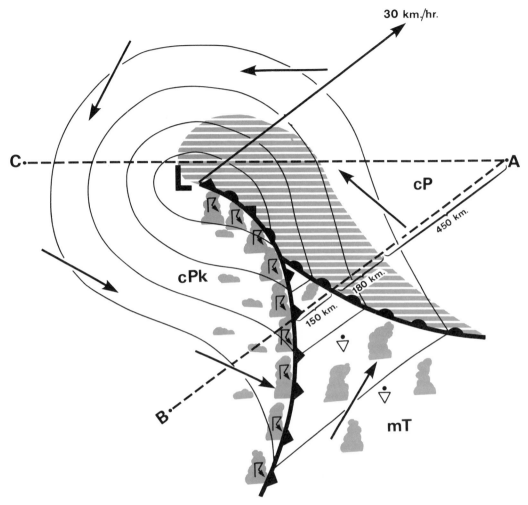

Figure 10.3. Partially occluded middle-latitude cyclone in the Northern Hemisphere.

rapidly clears in cold, fresh air. Since there is usually a k air mass behind the cold front, surface turbulence may produce small, flattish cumulus of fair weather for some distance behind the front—primarily daytime phenomena that dissipate at nightfall.

Since the warm air generally is moving away from the cold front and is only being underridden because the cold front is moving faster, no sustained flow of warm air continues at high levels above the cold front. In fact, on occasion the upper air, typically from some westerly direction, may be moving faster than the cold front itself, so that the air moves down the slope of the cold front rather than up it. Then there exists what is known as a *katabatic* front rather than an *anabatic* front, which is the more typical. On a significant number of occasions, a katabatic cold front may sweep through an area very quickly without any associated weather, the only indications of a frontal passage being a brief period of brisk winds and rapidly falling temperatures. In general, the more rapidly moving a front or a cyclonic system, the weaker the vertical development and the less intense the weather associated with it.

Figure 10.4. Vertical cross-section of fronts and weather along line AB in Figure 10.3.

Figure 10.3 illustrates that if we are situated at A and the storm center is moving in a northeasterly direction, as shown by the heavy arrow, at 30 kilometers per hour, a predictable sequence of events will take place. In about 15 hours the sky will become overcast and precipitation will begin. Continuous precipitation will last for six hours, during which the sky will be overcast with stratus clouds, the surface wind will be southeasterly, and the temperature will be cool. Then the weather will change fairly abruptly. The overcast will break up, and the clouds will change from stratus to cumulus with blue sky in between; the continuous rain will be replaced by occasional showers; and the temperature will rise perceptibly, with a wind shift from southeast to southwest. This type of weather will last for five hours while the warm sector of the storm passes. Next, the wind will become very gusty and shift abruptly from southwest to northwest. Severe thundershowers will probably occur for an hour or so, temperatures will fall precipitously, and they will plunge even more after the thunderstorms have passed as clear, cold weather moves in from the west. This sequence of events is illustrated by the vertical cross-section shown in Figure 10.4, which was taken along the line AB in Figure 10.3.

Six hours after the above forecast, a new weather map becomes available and shows that the storm center has now changed its movement from northeasterly to due east. Now the portion of the cyclone shown by the line AC (Figure 10.3) will pass through our position. The wind will gradually shift from southeasterly to northwesterly, but through north, rather than through south as predicted. The temperature will gradually become colder, but there will be no period of warm weather and no frontal passages. The sky will remain overcast and precipitation will occur for about twice as long a period as predicted. The new sequence of events is depicted by Figure 10.5. Our forecast is a complete bust! And the only change was the storm's direction of forward motion, by about 45°.

This could have been predicted by looking at the upper-air charts and noting that the storm path was approaching the crest of a ridge in the upper troposphere (Figure 10.6). As was explained in Chapter 6, circulation systems in the lower troposphere tend to be steered by flow in the upper troposphere, generally following the standing wave patterns in the circumpolar whirl from west to east. Figure 10.6 shows that successive positions of the center of the surface low conform closely to the western limb of the upper tropospheric ridge. As long as the storm is in this portion of the standing wave, it will move along a fairly straight path toward the northeast. But as it approaches the crest of the ridge it will turn toward the east and, later, toward the southeast. This change in direction will also cause the storm to weaken, as was explained in Chapter 6.

Waves appear to form on the polar and

Figure 10.5. Vertical cross-section along line AC in Figure 10.3.

arctic fronts in preferred areas of cyclogenesis (birth of cyclones), and resultant cyclones follow preferred paths across the face of the earth. These are shown for the Northern Hemisphere in a very general way for the two extreme months, January and July, in Figures 10.7 and 10.8.

In North America during winter, preferred areas of cyclogenesis lie just to the east of the Rocky Mountains in Alberta or Colorado or in east Texas and along the Atlantic Coast, especially north of Cape Hatteras in North Carolina. Wherever they originate, their routes eventually converge on the New England–Maritime Province area of northeastern United States and southeastern Canada, as they move into the North Atlantic to merge with the Icelandic Low, which is well developed at that time of year.

Farther east, cyclonic storms move out of the Icelandic Low eastward and southeastward to cross Europe in three preferred paths: (a) along the Arctic coast of Scandinavia and the Soviet Union, (b) up the Baltic to the Arctic coast of the Soviet Union, and (c) through the Mediterranean Sea after which some continue eastward into the Middle East, and some swing northeastward across the Soviet Union toward the gulf of the Ob River in western Siberia, where they converge with storms following the Arctic and Baltic routes. The

Ob Gulf plays much the same role as the Gulf of the St. Lawrence in North America, a convergence region of cyclone routes. East of the Ob, the storms usually die out along the Siberian fringes of the Arctic.

Few cyclonic storms penetrate the deep interior of Asia. But they form along Far Eastern segments of the polar front, usually off the Asian mainland south and east of Japan. They swing northeastward into the Aleutian Low and then farther eastward to penetrate the mountainous western portions of North America.

During July tracks are similar but are generally shifted poleward, and frequencies of storms along the routes are reduced. In Europe the Mediterranean route is largely abandoned. The Baltic cyclones penetrate Siberia as a broad, diffuse thermal low replaces the thermal high of winter. In the Far East the weakened polar front shifts westward across the Japanese islands and portions of the mainland, especially along the Mongolian-Soviet border, which becomes a significant area of cyclogenesis during mid- and late summer. Cyclones forming along the Mongolian Front move eastward across the Sea of Okhotsk into the northern Pacific.

It can be seen from these generalized maps that cyclones tend to edge their way poleward and converge in the subpolar low areas. Anticyclones (high pressure cells), on

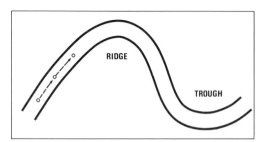

Figure 10.6. Small circles depict successive positions of surface low relative to wave pattern of upper tropospheric flow.

the other hand, in the higher middle latitudes work their way equatorward to join the subtropical high pressure belt (Figures 10.9 and 10.10).

Thunderstorms and Tornadoes

As was demonstrated in Chapter 8, thunderstorms may develop if the air is conditionally or convectionally unstable. For instability to be triggered, the air must be lifted to its level of free convection, either by heating at the surface, cyclonic lift over a front, or orographic lift (over a terrane barrier). Once the level of free convection has been reached, vertical movement is perpetuated by the release of latent heat of condensation, and vertical cells of circulation form that are somewhat self-contained thermodynamic mechanisms. These convective cells produce the commonly observed cauliflower-shaped cumulus clouds that may grow rapidly upward until lightning and thunder occur and a full-blown thunderstorm develops.

Figure 10.11 shows the progressive stages of the development of a thunderstorm. Initially, as the cumulus cloud is growing, the vertical air movement throughout the entire cloud is upward. To satisfy mass continuity, of course, an equal amount of downward motion must occur somewhere, and this takes place around the cloud formation. Thus, cumulus clouds are typically separated from each other by clear air where the downdrafts occur. The downdrafts are

usually much more diffuse than the updrafts, so the downward motion is not as noticeable. In its initial stages, each cumulus cell typically is no more than one kilometer in diameter, if that, while the space between cells is considerably greater.

As the cumulus cell grows well above the freezing level in the atmosphere and individual cloud droplets increase in size, the mature stage is reached and precipitation begins. This usually occurs first on the leading edge of the storm, according to its direction of motion. The falling precipitation produces a frictional drag on the air in that portion of the cloud and also cools the air significantly, both by contact and by partial evaporation of some of the falling precipitation which extracts the heat of evaporation from the surrounding air molecules. This results in a strong downdraft in the front edge of the cell. Usually the first noticeable characteristic of a thunderstorm's approach is a sudden shift in wind direction, accompanied by a blast of cold air descending from the roll cloud that is boiling along at the base of the leading edge of the cell, a short distance above the ground. This is known as a "gust front."

About this time the top of the cloud becomes glaciated (composed of ice crystals), and the typical "anvil" forms which is composed of milky-white cirrus clouds made up of ice crystals. This shape, which has been likened to the old-time blacksmith anvil, is formed as the upper tropospheric winds extend the glaciated portion of the cloud tops a long distance downwind underneath the tropopause inversion.

Now the cell has reached its advanced stage, and heavy precipitation throughout the cloud has produced downdrafts everywhere. This is the stage of heaviest precipitation, but it is also the beginning of the end, since no more warm, moist air is being brought in from below to release its latent heat to keep the convection going. In the course of ten or fifteen minutes the bulk of the precipitation falls, and the

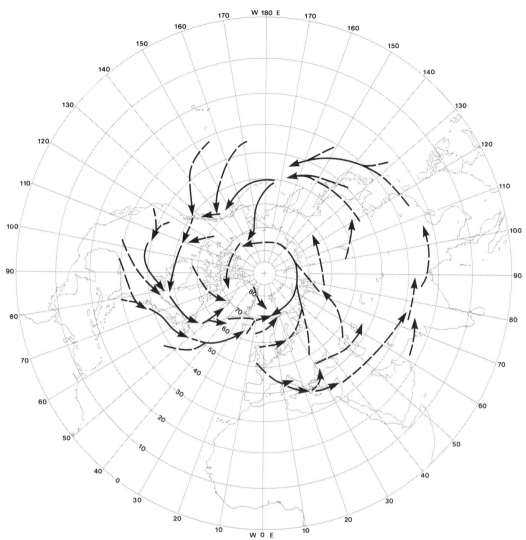

Figure 10.7. Cyclone paths in the Northern Hemisphere, January. Solid arrows denote primary tracks; dashed arrows, secondary ones. (After William H. Klein, Principal Tracks and Mean Frequencies of Cyclones and Anticyclones in the Northern Hemisphere, *U.S. Weather Bureau, Research Paper No. 40 [Washington, 1957].)*

convection cell slowly dissipates as other cells grow around it. The precipitation time-span and the intensity of the cell may be greatly enhanced if the convection cell is tilted so the precipitation at the leading edge does not fall through the updraft in the main body of the cell. The most severe thunderstorms often develop a double-vortex structure, counterclockwise in their equatorward portion and clockwise in their poleward portion.

Thunderstorms are by far the most numerous of the violent, turbulent storms. It has been estimated that more than 40,000 of them take place on earth every day. Since their development depends upon conditionally unstable air and high humidity content, most are found in association with maritime tropical (mT) air masses. As can be seen in Figure 10.12, thunderstorms occur most frequently in the equatorial region, particularly on the continent of

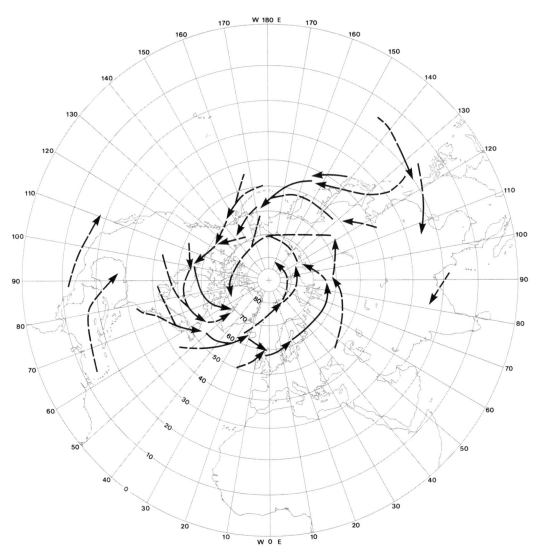

Figure 10.8. Cyclone paths in the Northern Hemisphere, July. (After Klein, Cyclones and Anticyclones in the Northern Hemisphere.*)*

Africa. Within the United States they occur most frequently in the Southeast, which is dominated by warm, moist, conditionally unstable air along the western edge of the North Atlantic High.

Although tropical and subtropical thunderstorms probably produce the greatest amounts of precipitation, they are not necessarily the most turbulent. Extremely turbulent thunderstorms, attested to by large hail and occasional tornadoes, are found in portions of the middle latitudes, particularly in central United States and north-central India. Such turbulence is usually related to a rapid decrease of absolute moisture content with height, which sets the stage for convective instability, triggered either by orographic or cyclonic uplift and strong vertical wind shears.

Individual thunderstorm cells represent

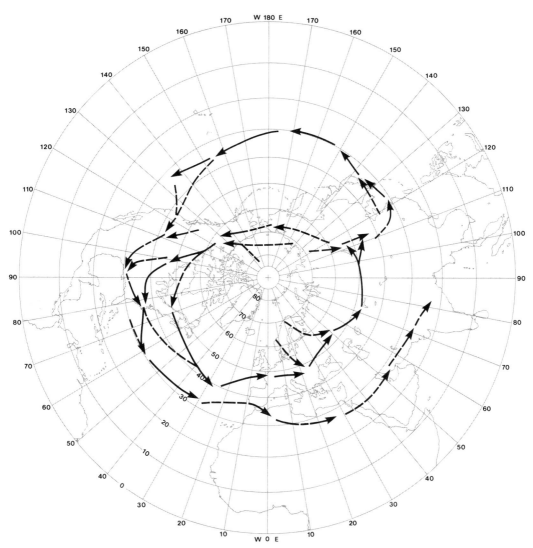

Figure 10.9. Anticyclone paths in the Northern Hemisphere, January. (After Klein, Cyclones and Anticyclones in the Northern Hemisphere.)

small circulation systems in the atmosphere that are usually imbedded in broader-scale systems, such as a cyclonic storm, the intertropical convergence zone, or a conditionally unstable k air mass, such as the maritime tropical air moving from the Gulf of Mexico into southeastern United States during summer. Individually, they cover only a few kilometers in diameter at best, but frequently they coalesce into thunder-

storm masses consisting of several cells, and occasionally they form lines of almost continuous thunderstorm cells that move in unison as squall lines.

Squall Lines

In central United States squall lines sometimes form in warm sectors of cyclonic storms 50–300 kilometers ahead of cold

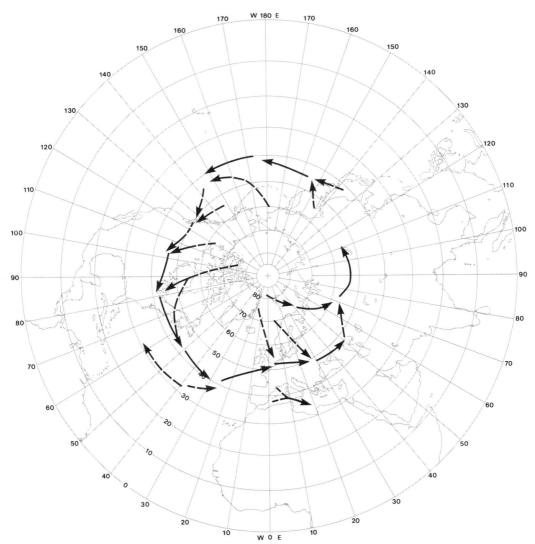

Figure 10.10. Anticyclone paths in the Northern Hemisphere, July. (After Klein, Cyclones and Anticyclones in the Northern Hemisphere.*)*

fronts (Figure 10.13). Downdrafts in the leading edges of thunderstorm cells associated with cold fronts appear to give rise to small high pressure centers that push air (cooled by evaporation or precipitation) ahead of cold fronts in the form of pseudofronts, which underride warm air ahead and trigger instability. Squall lines in this region also appear to be related to low-level jets of moist air blowing from south to north across the Great Plains, particularly from Texas to Nebraska. Similar occurrences are found in east-central Argentina.

Squall lines move with about the same speed and direction as the cold fronts behind them, and their approach is often mistaken as the approach of the cold front, which confuses weather predictions. In some cases squall lines move faster than the cold

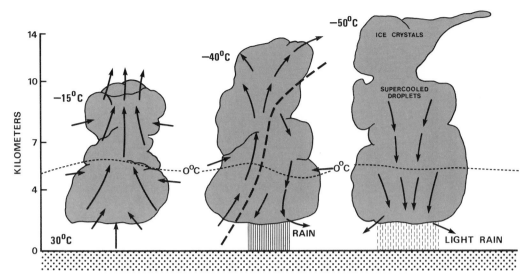

Figure 10.11. Stages in the development of a thunderstorm cell. Left: *cumulus or developing stage.* Center: *mature stage.* Right: *old or dissipating stage.*

fronts behind them and eventually are no longer directly associated with the cold fronts. One of the most remarkable occurrences of squall lines is in west Africa in the Lake Chad–Niger River area, where they become the primary cause of rainfall during summer. Here they occur in lower latitudes unassociated with the polar front and its cyclonic storms. Although they are imbedded at the surface in a southwest monsoonal flow, they move westward with the winds aloft that flow from east to west.

Tornadoes

Squall lines that form near the apexes of warm sectors in cyclonic storms provide some of the most turbulent weather in interior North America and on extreme occasions may spawn tornadoes, the most violent of all weather phenomena. The characteristic funnel cloud builds downward from low-hanging cumulonimbus clouds associated with severe thunderstorm cells and forms a very small vortex, typically less than 400 meters in diameter, which produces extreme wind speeds, perhaps as

much as 225 meters per second (500 mph). At the surface the air rotating around the vortex moves inward toward the center and then upward, carrying with it debris as large as automobiles and major portions of houses. Extreme pressure gradients are produced over short distances, perhaps as much as 250-millibars change in a few hundred meters. These add to the destruction, since tightly closed buildings may not allow air inside to escape fast enough to equalize the pressure within the building and the pressure dropping outside with the passage of the storm. Tornadoes often leave paths of destruction in urban areas where entire blocks of houses appear to have literally exploded outward, with their roofs lifted off and walls laid out from the centers of the buildings.

A typical tornado in the United States moves northeastward with a forward speed of perhaps 20 meters per second (45 mph). It frequently skips, with the funnel cloud temporarily receding upward into the base of the cumulonimbus cloud and later dropping down again. The fleeting lifespan and erratic movement make it difficult to trace

Figure 10.12. Average annual number of days with thunderstorms. (From Thunderstorms, *vol. 2 [NOAA, 1982], p. 34.)*

the path of any single tornado vortex. Although tornadoes are nearly totally destructive where they hit, their routes are narrow and discontinuous, and their appearances do not significantly alter average statistics of winds and other climatic phenomena, even in the regions of their most frequent occurrence. But if they do hit, they are phenomena to be reckoned with. In an attempt to predict their occurrence and warn people in affected areas, the National Weather Service operates a severe-storm warning center in Kansas City, Missouri, with a laboratory near Oklahoma City, near the center of most frequent tornado occurrence.

Most of the tornadoes on earth occur in the United States east of the Rocky Mountains. Although all states have recorded tornadoes, the maximum number occur in the southern and central Great Plains, the Midwest, and the southeastern United States. Outside North America, Australia experiences tornadoes most frequently, particularly in the southeastern part of the continent. Only scattered occurrences of tornadoes have been reported in other parts of the world.

In central and southern United States tornadoes are most likely to develop near stalled sections of the polar front during periods of most intense air mass contrasts, which usually occur during spring. In the states of maximum occurrence, Oklahoma and Kansas, three-fourths of the tornadoes

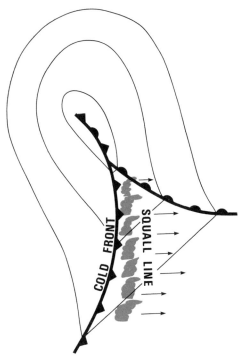

Figure 10.13. Typical squall line ahead of cold front in warm sector of cyclone in central United States.

take place during the months of April, May, and June. Most occur during late afternoon and evening during the period of maximum diurnal convective activity. In southeastern United States, particularly along the Gulf Coast, multiple small tornadoes are commonly associated with hurricanes during late summer and autumn.

11 | FLOW PERTURBATIONS IN LOWER LATITUDES

In lower latitudes, because of reduced temperature and pressure gradients and reduced Coriolis Force, the geostrophic wind concept breaks down, and the air tends to move more and more perpendicular to the isobars as the equator is approached. As a consequence, perturbations in the air flow that may bring about notable weather take on different forms than they do in middle and higher latitudes.

The Trade Wind Inversion

As mentioned earlier, the air circulating around a subtropical high pressure cell undergoes divergence and subsidence, which is strongest on the equatorward-eastern edge of the cell. Thus, a strong subsidence inversion is formed not far above the surface along western coasts of continents in subtropical latitudes, such as southern and Baja California, northern Chile and Peru, and northwest and southwest Africa. As the air turns westward and moves into progressively lower latitudes along the equatorward margins of these highs, the surface air picks up heat and moisture from the tropical sea, which produces increasing instability in the marine layer underneath the inversion and lifts the inversion layer to greater and greater heights. For instance, during summer on the west coast of North America near San Francisco, the height of the base of the inversion averages about 400 meters above sea level, while at Hawaii 4500 kilometers to the southwest, it averages more than 2000 meters above sea level. Farther west the inversion disappears altogether as the air becomes conditionally unstable along the western edge of the high.

In the equatorward-eastern quadrants of the subtropical highs where subsidence is strongest and the inversion base is lowest above the surface, the marine layer of air under the inversion is usually too thin for convection to produce clouds of significant vertical development. In these places fog and low overcast stratus clouds are the rule. If the inversion is very low above the surface, the marine layer may not be thick enough even to produce stratus clouds, as is often the case in southern California during middle and late summer. But as the inversion lifts westward, convective activity within the marine layer typically produces cumulus clouds that build up to the base of the inversion. These are the well-known "trade wind cumulus" that dominate the eastern halves of subtropical ocean areas. They are very regular formations when viewed from above, such as from the high volcanic peaks on the Big Island of Hawaii, where one can observe them moving in from the northeast, their bases regularly positioned at the condensation level a few hundred meters above the surface of the sea, and their tops distinctly limited by the base of the inversion, perhaps 2000–2500

Figure 11.1. Trade wind cumulus moving with the northeast trades onto the Big Island of Hawaii below the summits of Mauna Loa (distance) and Mauna Kea (foreground). Note the barren surface of the dry summit of Mauna Kea.

meters above the sea (Figure 11.1). From such cloud formations, intermediate slopes of windward sides of mountains may receive copious precipitation, upward of 7500 millimeters (300 inches) per year or more, while lower and higher slopes remain dry. It has been estimated that the top of Mauna Kea on Hawaii at an elevation of 4206 meters (13,796 feet) receives no more than 375 millimeters (15 inches) of precipitation per year because it is above most of the clouds. Similar conditions exist at Tenerife in the Canary Islands off the northwest coast of Africa.

Easterly Waves

Along the equatorward-western edges of the subtropical highs imbedded in the conditionally unstable trade winds appear open wave formations, 2000 kilometers or more in length, that move westward with the trades at forward speeds averaging about 5 meters per second (11 mph). Therefore, an entire wave takes four or five days to pass a position. Although these are very subtle features in the surface isobaric pattern, which is usually drawn with little available data anyway, they do impart an aperiodic element to an otherwise rather regular diurnal cycle of weather events in the tropics (Figure 11.2). The easterly waves are associated with divergence, subsidence, and clear skies on their leading edges, followed by convergence, uplift, and increased thunderstorm activity after their wave crests pass. Thus, the passing of an entire easterly wave over a region may result in two to three days of less than normal clouds and precipitation, followed by two to three days of increased thunderstorm activity and precipitation. The waves show no tendency to go through a cycle of occlusion, dissipation, and regeneration, as they do in the middle latitudes. Little is known about the origin and fate of these minor perturbations in the trade wind flows.

Hurricanes

Although pressure gradients and winds are usually weak and relatively unorganized in the tropics, occasionally rotary motions are initiated which, under the influence of a weak Coriolis parameter, develop into high wind speeds. Since these disturbances originate in tropical oceanic areas where few surface observations exist, little is known about the initiation of the cellular circulations with pronounced low pressure centers. Satellite photography during the last couple of decades has helped to locate these storms earlier in their lifespans, but still they are not observed until they are fairly well developed. Some of them may originate as easterly wave perturbations or open wave formations on the intertropical front, where it has migrated a considerable distance from the equator. These storms do not occur right on the equator, presumably because there is no Coriolis Force to initiate the rotary motion. Most of them seem to originate 5°–10° latitude on either side of the equator and move westward with the trade winds, gradually working their way poleward until they come under the influence of the westerlies and then curve into middle latitudes. Figure 11.3 shows typical routes of travel of tropical storms that have developed to hurricane strength.

The driving energy seems to be primarily the latent heat of condensation. These storms develop only over ocean areas that have temperatures higher than 27°C (80°F). Air converging toward a low pressure center over a warm water surface carries great amounts of energy in the form of latent heat tied up in the moisture that has been evaporated from a broad ocean surface and concentrated in a small area. Rising motion around the center of the low cools the air adiabatically and releases this energy through the process of condensation. As long as the moisture source, and hence the energy source, is available, the storm con-

Figure 11.2. Model of an easterly wave in the Caribbean. Solid lines represent isobars at the surface; dashed lines represent streamlines at the 500 mb. level. The heavy line marks the trough line. Stippling indicates area of increased precipitation.

tinues to develop. But when the storm leaves the warm ocean surface, either by moving into higher latitudes or over a continental mainland, it quickly loses the infeed of new energy, and the wind speed dies down almost immediately. The low pressure center with its associated widespread cloudiness and heavy precipitation may move far inland, but the strong winds generally are limited to within 30–40 kilometers of a seacoast. For instance, in the United States a hurricane storm may come onto the mainland from the Gulf of Mexico, move up the Mississippi Valley, curve northeastward up the Ohio Valley into the New York or New England area, and drop several hundred millimeters of rain over huge areas, thereby causing widespread damage from flooding. But local inhabitants are not aware that they are undergoing a hurricane passage, because the wind speeds are no higher than they are with ordinary extratropical cyclonic storms common to the region. In fact, such storms frequently join up with extratropical cyclones coming across the central United States, which causes them to deepen and stagnate, thereby prolonging their influence over the northeastern states. The amount of precipitation released in such a situation can be phenomenal. Hurricane Agnes, in June 1972, dumped about 300 millimeters (one foot) of rain over much of eastern Pennsylvania and adjoining states, causing tremendous

Figure 11.3. Characteristic tracks of tropical storms. Shaded areas have mean sea-surface temperatures exceeding 27°C during September in the Northern Hemisphere and March in the Southern Hemisphere.

1500 Miles

amounts of damage from flooding and mudslides.

In coastal areas, in addition to the high winds, one of the most damaging characteristics of a hurricane storm is the surge of wind-driven ocean water as the storm approaches the coast. It is not uncommon for the water level to rise more than 6 meters (20 feet), which, along flat coastal plains, may drive the water far inland. During the night of 17–18 August 1969, Hurricane Camille drove a 7-meter tide ahead of it across the Louisiana-Mississippi coast and literally drowned the small town of Cameron, Louisiana. Such occurrences have been particularly disastrous in the densely populated Ganges delta at the head of the Bay of Bengal, where such storms repeatedly have caused hundreds of thousands of deaths by drowning and untold property damage. The most memorable case, on 12 November 1970, had an official death toll exceeding 200,000 people and, as a result of its millions of homeless refugees, eventually touched off the Pakistani civil war that resulted in the establishment of the new country of Bangladesh.

According to official definitions of the World Meteorological Organization, closed cyclonic circulations in the tropics with wind speeds less than 17 meters per second (38 mph) are tropical depressions. When the wind speed rises to 17–32 meters per second they are called tropical storms. When wind speeds reach more than 32 meters per second (75 mph) they are called something else, depending on the region. In the Caribbean area they are known as hurricanes and in the Pacific, generally, as typhoons, but in different portions of the Pacific they are known by different names. In the Philippines they are known as *baguios*, after a small resort city that in 1911 received 1150 millimeters (46 inches) of rain in 24 hours during the passage of such a storm. In the Indian Ocean area, including the Bay of Bengal and the Arabian Sea on either side of the Indian peninsula, they

are known as cyclones, as they commonly are in Australia. Hence, once again the term "cyclone" is used in a different context. They are all the same kind of storm, although usually of different intensities in different areas.

These storms are most frequent and most intense in the western North Pacific, where wind speeds of 70–90 meters per second (150–200 mph) are not uncommon, although the majority of them have wind speeds somewhat less than that. During the height of their season (July–October) it is not uncommon to find two, or even three, such storms in different stages of development following one another across the western part of the North Pacific, 1500–3000 kilometers apart.

The second area of most frequent and most intense occurrence of these tropical storms is the western part of the North Atlantic and adjacent Caribbean Sea and Gulf of Mexico. Here wind speeds very rarely, perhaps only once in 10 or 20 years, reach 90 meters per second (200 mph). Other regions in descending order of occurrence are the northern Indian Ocean, particularly the Bay of Bengal, but to some extent the Arabian Sea; the southern Indian Ocean in its western portion next to Madagascar Island off southeastern Africa; the western South Pacific, particularly to the east of Australia, but also all along the northern coast; and the very eastern North Pacific southwest of Mexico, which has fewer and usually weaker storms. The outstanding exception to such locations is the western south Atlantic next to Brazil, which has never experienced hurricane-type storms (see Figure 11.3). The sea-surface temperature there is usually too cool for the formation of such storms, and the position and shape of the south American continent are such that this coastal area juts well into the central portions of the South Atlantic high pressure cell, so that atmospheric stability is generally greater in this area than in the others mentioned.

In the Northern Hemisphere, highest sea surface temperatures are reached during June–November, and these are the months in which hurricanes develop most frequently, although occasionally they occur in adjacent months of the year. In the north Indian Ocean, which is consistently warm, they have occurred every month of the year. In the Southern Hemisphere, sea-surface temperatures are highest from December through March, and that is the season of most frequent tropical storms.

A fully mature hurricane may range in diameter from as small as 100 kilometers to more than 1500 kilometers. Wind speeds increase from the periphery toward the center, where sea-level pressure may reduce to less than 900 millibars. The lowest sea-level pressures ever recorded are in the centers of such storms. (Pressure may reach even lower levels in the centers of tornadoes, but no measurements or adequate estimates have been possible in those cases.) The official world's lowest sea-level air pressure was recorded in the eye of Typhoon June in the western North Pacific in November 1975, where the central pressure dropped to 25.90 inches (876 mb).

In the immediate center of the storm the wind dies down, and it becomes relatively calm and clear in what is known as the eye of the storm. It is obvious that there is in the center of the hurricane a chimney of descending air that produces subsidence and relatively clear weather, accompanied by only light winds. This may cover a circular area approximately 15–45 kilometers in diameter. A distinctive wall of clouds surrounds this eye, and in this portion of the storm the greatest uplift, highest wind speeds, and maximum precipitation occur.

Around the eye wall the clouds arrange themselves in spiral bands, as shown in the satellite photo in Figure 11.4. Throughout much of the storm there appear to be alternating spiral bands of uplift and subsidence, the bands of uplift being marked by the bands of clouds, and the bands of subsidence by the relatively clear areas in between. General subsidence appears around the periphery of the storm, particularly on its leading edge, which has prompted the old adage "the calm before the storm." Often during an entire day as a hurricane approaches, the weather is unusually calm, clear, and sunny.

Hurricanes frequently extend up to the tropopause. Because of the subsidence and accompanying adiabatic heating in the eye, the center of the storm remains warmer than its surroundings right up to the tropopause. Thus, the hurricane is a warm-core cyclone and must decrease in intensity with height, eventually becoming an anticyclone aloft.

Equatorial Westerlies

Between the trade winds, a belt of surface westerly winds is often found near the equator, especially in the summer hemisphere. The most extensive zone stretches from just west of Africa eastward into the western Pacific. A limited zone intersects the equatorial west coast of south America. Here and in equatorial western Africa, westerly winds prevail throughout the year. But in eastern Africa, southeast Asia, and northern Australia, they shift latitudes widely with the seasons. Their most poleward extent is found during the Northern Hemisphere summer in northwestern India-Pakistan, where they sometimes penetrate to 30°N latitude as the Indian monsoon. Over the Indian Ocean they often extend upward to 5–6 kilometers, but in the rest of the world they tend to be shallower. Over Africa they generally extend only to 2–3 kilometers height. The westerlies have a meridional component directed poleward and tend to be more unstable than the tropical easterlies with equatorward components. The westerlies usually bring heavy rainfall to areas consistently affected by them, particularly along mountainous windward coasts, such

Figure 11.4. Satellite photo of Hurricane Allen in the western Gulf of Mexico, 8 August 1980. Note the spiral cloud bands and the eye.

as the mid-section of the western coast of Colombia in South America, portions of the Guinea-Cameroon coast of Africa, and the West Ghats and Khasi Hills of India.

The Southeast Asian Monsoon

One of the most outstanding features of the air flow in low latitudes is the great seasonal reversal of winds and resultant weather that occurs throughout much of the north Indian Ocean, India, the countries of southeastern Asia, much of China, and parts of adjacent countries, as well as the western fringes of the North Pacific Ocean.

To some extent it also affects much of Indonesia, where it extends south of the equator into northern Australia. Discussion here will be confined to its most classic development over the Indian subcontinent and adjacent water bodies—the Arabian Sea to the west and the Bay of Bengal to the east.

Here, particularly, its specific features appear to be due to much more than just the seasonal reversal of land-sea heating differentials. The suddenness of the "burst" of the rainy period during early June in southwestern India, the deflection of the southwest surface winds at the head of the Bay of Bengal to southeasterlies up the

Ganges Valley, and the progression of "monsoon depressions" from the Bay of Bengal northwestward across India are some of the features imbedded in the general summer monsoon environment that cannot be explained by simple land-sea temperature differences.

The highlands to the north appear to play a major role. The massive Tibet Plateau, ranging in average elevation from approximately 4000 meters in its eastern portions to 6000 meters in its western portions, with still-higher surmounting and surrounding mountain ranges, pokes well into the upper troposphere to split the westerly air streams of the east Asian region. During winter two jet streams occupy positions to the south and north of this highland mass. The southerly one down the Ganges Valley of northern India is by far the stronger of the two. But in spring it begins to weaken, and by late May or early June it disappears altogether or flops over the highlands to join the jet on the northern side. This event allows for the "burst" of the summer monsoon rains over much of peninsular India, associated with southwesterly airflow at the surface.

An easterly jet then develops near the 150-millibar level in the upper troposphere which, during summer, is centered at about 15°N latitude and spreads from central India eastward and westward to eventually occupy an extensive area all the way from the western Pacific westward through the northern portions of the Indian Ocean, the southwestern tip of Arabia, and across north Africa to the eastern Atlantic Ocean. This appears to stem, partially at least, from the formation of a thermal high pressure cell in the mid- and upper troposphere over Tibet and surrounding highlands. This upland surface absorbs much heat from the strong insolation coming through the thin, clear air at this altitude, which produces temperatures in the air above the highlands much warmer than they would be at similar altitudes in the free atmosphere without an insolation-absorbing surface. Much heat energy is also injected into the upper troposphere along the southeastern fringes of Tibet by the release of latent heat from copious precipitation occurring along the southern mountain slopes in northeastern India at this time of year. The warm air temperatures create a thermal high above the highlands which, rotating clockwise, brings on the formation of an easterly jet along its southern flank.

The easterly air flow aloft over India deflects the southwesterly surface winds at the head of the Bay of Bengal and causes the surface air to proceed up the Ganges Valley from southeast to northwest. It also guides the "monsoon depressions," weak enclosed lows originating in the Bay of Bengal area, northwestward across India.

What appears to be a segment of the intertropical front becomes positioned far north during the summer in the Ganges-Indus Plain of northern India-Pakistan. But in the Indus Valley it does not bring on precipitation. The extreme dryness of the Thar Desert in this region and the rap.. decrease of precipitation northward along the west coast of peninsular India are apparently explained by the abnormal slope of the frontal surface in this region—upward to the south—as the hot desert air from the north overrides the less hot marine air from the south off the Arabian Sea. The frontal surface acts as a lid and limits cumulus buildup in the marine air underneath. Only when the marine layer becomes thick enough, far south of the surface front, do showers commence, usually off the coast well to the south of Pakistan.

During winter the surface air generally flows from northeast to southwest across much of India. The westerly jet becomes reestablished over the Ganges Plain and brings in some weak surface lows from the Middle East, which produce small but critically beneficial amounts of rain to Pakistan and northwestern India upon which the winter wheat of this area depends.

12 | WIND

In this chapter wind will be considered as a climatic element, as it influences objects at the earth's surface. It is usually perceived as a significant climatic element only when it is persistent enough or strong enough to be annoying or damaging, or distinctive enough to be a characteristic part of the climate of an area.

Climatic Representation of Winds

Although surface winds are measured fairly accurately and consistently every hour, as are precipitation and temperature, the wind is more difficult to deal with statistically since it is a vector consisting of two components, direction and speed. The most complete way to depict winds at a place over a period of time is in devices such as frequency tables or wind roses, showing the percentage of all time when the wind blows from each direction of a 16-point compass. These usually represent only direction, although average speed from each direction can be added. In any case, such representations are bulky and difficult to compare among different stations.

Prevailing Wind

One of the easiest climatic expressions of wind is the so-called "prevailing" wind that represents all winds at a point by the direction from which they blow most fre-

quently. But if the winds are variable, the prevailing wind might not be particularly representative, since it might blow less than the majority of the time. Figures 6.2 and 6.3 illustrate prevailing winds of the world by directions of arrows. The widths of the arrows indicate percentages of all time that the winds blow from the prevailing directions and thus show how prevailing the prevailing winds are. It is immediately obvious from these maps that the winds are most consistent in direction in the trade wind belts on either side of the meteorological equator, and probably least consistent over land masses in middle latitudes. These maps indicate nothing about wind speeds, which could be depicted by varying the lengths of arrows. More commonly, direction, constancy, and speed are depicted by varying direction, length, and width of arrows, respectively.

Resultant Wind

Perhaps more representative of the total air movement at a given point is the so-called "resultant" wind. This is the vector addition of all the winds at a given point over a period of time, such as a month, for which the winds are being depicted. Theoretically, this represents the net movement of air during the period. Resultant winds can be determined graphically by connecting heads and tails of arrows whose directions and lengths represent the directions and speeds of individual wind observations. Or resultant winds can be derived statistically by dividing each observed wind direction and speed into north-south and east-west components, averaging all these components, and then recombining them into a resultant wind vector.

As mentioned, the resultant wind theoretically represents the net motion of air across the point over a period of time. Consequently, it might not represent any wind that blows, and in a region of variable winds might significantly underrepresent the windiness of the area. To take an oversimplified case as an example, if the wind in a mountain valley is confined to blowing either up or down the valley and does both with equal frequency and equal speed, the resultant wind is zero. Yet calm conditions may never occur in the valley, and in fact rather strong winds might be experienced much of the time.

Where winds blow equally frequently from opposite directions, a prevailing wind cannot be designated, either. In such cases, as far as influences of the wind on surface objects are concerned, it might be more useful simply to average wind speeds and forget about directions. Wind direction is very significant for weather analysis and forecasting, but perhaps wind speed is more significant as far as effects on external objects are concerned.

Resultant wind arrows have been utilized in so-called *streamline analyses.* Individual arrows computed for resultant winds are connected by smoothly curved lines that theoretically depict net air flow over broad parts of the earth's surface and clearly reveal lines and vortices of convergence and divergence, as shown in Figures 6.8 and 6.9. Such streamlines do not relate directly to either constancy of direction or speed of the winds. In mountainous areas it is unrealistic to draw streamlines that ignore details of the topography.

Characteristic Wind

In some areas, occasional characteristic winds, while too infrequent to alter prevailing or resultant wind determinations, may nevertheless have profound effects on the inhabitants of the regions. Examples are the foehn winds of southern California and many other mountainous areas of the world, which blow only occasionally but can bring on the highest temperatures, lowest humidities, clearest skies, and greatest fire hazards of the year. Other examples are extreme winds associated with occasional storms, such as tornadoes and hurricanes, which have already been discussed in Chapters 10 and 11. These phenomena have such profound effects that, although often too infrequent to alter significantly long-term averages of wind directions or speeds, they must be taken into account in the climate of a region. Thus, in a full climatic description of an area one must include the characteristic winds determined by inspection of day-to-day records rather than long-term averages.

Local Winds

Characteristic of many areas are local winds imposed upon the general circulation by topographic influences. Most common are sea and land breezes, mountain and valley breezes, foehns, bora, and katabatic winds.

Sea-Land Breezes

Sea and land breezes tend to set up in a diurnal regime along the coast of any major water body because of different temperature reactions to heat inputs and losses. During daylight hours the air over land warms more rapidly than over adjacent water and eventually becomes warmer and lighter than the air over the sea. A vertical circulation cell is set up, with sea air moving inland at the surface and land air moving seaward aloft. This so-called "sea breeze" will usually begin blowing inland late in the morning and continue through the afternoon. Such breezes may become particularly strong, 10 meters per second (22 mph) or more, along coasts paralleled by cold ocean currents, such as the coast of southern California or the Peruvian coast of South America. In such areas they may penetrate 30 or more kilometers inland and in some cases considerably more, if the inland topography contributes to them. At night the land cools more rapidly than the sea, and after midnight the air over the land becomes colder than that over the sea, so a surface land breeze is set up. This is usually less well developed than the sea breeze during the day.

Along subtropical coasts paralleled by cool ocean currents, the land-sea breeze regime may be a diurnal occurrence practically every day of the year and become the dominant wind feature in the immediate shore area. On other ocean coasts it may occur only occasionally, when the general circulation over the area is weak. Large lakes, such as the Great Lakes of North America, may also cause diurnal regimes, but they are usually much less pronounced and penetrate no more than a few kilometers inland.

Monsoons

In many coastal areas of the world, land-sea exchanges of air may be set up on a seasonal basis. Such seasonal reversals of winds are known as monsoons. By far the best known are the monsoons of southeastern Asia, but as was mentioned in Chapter 11, the seasonal reversal of air flow there is due to more than just seasonal reversals in the heating of land and sea. Conforming more closely to a seasonal reversal of wind caused by land-sea temperature differences is the Arctic coast of Siberia. During winter, cold winds blow from south to north off the cold land toward the relatively warmer air over the Arctic

ice. Even though the Arctic Ocean is frozen over right up to the coast during winter, considerable heat is conducted upward from the unfrozen subsurface water through the approximately three-meter-thick surface ice to maintain surface air temperatures around $-35°C$, which is usually 10 to 15 degrees warmer than over the Siberian landmass to the south. During summer, winds blow from north to south from the relatively cold fringes of the Arctic seas onto the warmer landmass. Ice floes keep water temperatures only three or four degrees above freezing during summer, while surface air temperatures over the land average $10°-15°C$.

Mountain-Valley Breezes

In mountainous topography, diurnal regimes of wind may be set up (much as they are along coasts) by the differential heating between valley floors and mountain slopes. During the day, convection currents from heated valley floors move up mountain slopes in the form of valley breezes. During the night, upper slopes of mountains cool rapidly due to radiational heat loss through thin air, and the cooled air moves downslope under the force of gravity to settle into valley floors as mountain breezes. Mountain breezes at night are generally the best developed of this regime, and they may become quite brisk if funneled through constricted areas such as mouths of canyons.

Where mountainous topography and coastlines are in juxtaposition, the two regimes can combine to pull sea air far inland during the day. This is particularly conspicuous in areas such as southern California, where air that has crossed the Pacific Coast at Santa Monica in the morning may be found in the afternoon at Palm Springs, 130 kilometers (80 miles) inland. Pollution control agencies in southern California, tracing the movement of individual particles injected into the air, have found that this much movement does indeed occur during the day. Of course, by the time it has reached Palm Springs the air has warmed to as much as 40°C (104°F) or more, and therefore has lost the characteristic of a sea breeze.

Foehns and Boras

In mountainous areas, other types of local winds may occur that are related to the general circulation over the region rather than to local heating differences. Such are *foehn* and *bora* winds—downslope winds on lee sides of mountains that are crossed by the general circulation of air over the region. These descending air flows warm adiabatically at the rate of 10°C per kilometer (5.5°F per thousand feet). If the air is not too cold to begin with, it arrives at the base of mountains as hot, exceedingly dry, clear air that can increase surface temperatures as much as 15° or 20°C in a few minutes and drop the relative humidity to 10 percent or less, thus creating extreme fire hazards. Temperatures are warmer than they were at the same elevations on the windward slopes because latent heat of condensation was added as the air dropped precipitation as it surmounted the windward slopes.

If the air is funneled through constrictions such as canyons, these winds can reach velocities of as much as 45 meters per second (100 mph) or more, but in open areas the air flow is generally mild, and the increase in temperature and drop in relative humidity are the most discernible characteristics. Such conditions are known as "foehn" winds, a German term meaning a fall or descending wind down a mountain slope. Foehn winds occur in many mountainous areas of the world, sometimes on a seasonal basis if the general circulation of air has a seasonal character. During winter they can be very beneficial because they can cause snow covers to evaporate (sublime), thus keeping pastures open for grazing. The nomadic herders of Central Asia have a saying that "One day of foehn is worth two weeks of sunshine." On the other hand, foehns may induce rapid thawing of snow and disastrous avalanches in mountainous regions, such as the Alps and Caucasus. In some regions effects are so spectacular that foehns are given special names, such as the "Santa Ana" of southern California and the "Chinook" of the northern Great Plains along the eastern foothills of the Rocky Mountains.

If the descending air originally is much colder at the tops of the mountains than the air normally is at the bases of lee sides, the air may arrive at low elevations colder than the normal temperatures, even though the air has warmed adiabatically during its descent. Such is the case during winter when cold air builds up over the Swiss Plateau of Europe and then descends the southern slopes of the Alps through several structural valleys. It frequently arrives on the Mediterranean and Adriatic coasts of France, Italy, and Yugoslavia as relatively cold winds with speeds as high as 20–30 meters per second (45–70 mph). This is very disturbing to vacationers along the Riviera at this time of year. Such winds have been termed "bora" winds, a term derived from the region around the head of the Adriatic Sea. In some areas such winds are known by local names, such as the "mistral" in southern France.

Katabatic Winds

Another kind of descending wind is the katabatic wind, which is caused simply by the force of gravity, cold air descending under its own weight. This is common in areas such as the fringes of the Antarctic and Greenland ice caps, where cold air from the high ice plateaus constantly slides downslope to the sea. These winds are typically gusty, since after they flow for a short while the cold air on the uplands becomes exhausted. After it rebuilds, more air slides down again. Therefore, in the fringes of these ice caps extreme blizzard conditions may occur where first the wind is relatively calm and then suddenly may blow with a speed of 25 meters per second or more, often changing direction abruptly.

Extreme Wind Speeds

Since heat and easterly momentum must be transferred from low latitudes to high latitudes with greatest flux through the middle latitudes, the atmospheric circulation by necessity is generally most vigorous in the middle latitudes. Thus, average wind speeds are generally higher in the middle latitudes than at either low or high latitudes. Of course, there are exceptions under certain conditions, such as the large temperature contrasts between land and sea along some high-latitudes coasts or on occasions when significant pressure gradients occur in lower latitudes. Remember that the Coriolis Force is very weak in low latitudes, and therefore even a weak pressure gradient can cause excessive winds, such as those found in hurricanes. The highest wind speeds on earth are found in tornadoes, which are largely middle-latitude phenomena, primarily in the United States east of the Rocky Mountains. Here it has been estimated that speeds as high as 225 meters per second (500 mph) may occur, although no one has stood there with a three-cup anemometer and measured them.

The highest measured surface wind speeds on earth have been recorded at an elevation of 1917 meters (6,288 feet) on the top of Mount Washington in New Hampshire, where the wind has reached a speed of 103 meters per second (231 mph).

There it has averaged 77 meters per second (173 mph) for one hour's duration and 58 meters per second (129 mph) during an entire day. Over the course of the year the wind there averages 16 meters per second (35 mph). This is somewhat less than an annual average of 19 meters per second (43 mph) at Cape Denison along the fringe of Antarctica. During July 1913, winds there averaged 25 meters per second (55 mph) over the entire month. Such winds and low temperatures produce extreme wind chills. The Soviets have estimated wind chill equivalent temperatures as low as $-150°C$ along the Arctic coast of eastern Siberia during winter, when strong land breezes are the rule.

Surface features, primarily vegetation or the lack thereof, greatly influence average wind conditions. Dry lands with meager vegetation and high surface temperatures are conducive to constant winds which, although they may not be extreme in speed, are often so persistent as to become exhausting. Similar conditions prevail over tundra vegetation in the treeless fringes of the Arctic and mountains above the tree line. In many of these areas the convective activity is strong during daylight hours, and there is often a diurnal regime to wind speeds. They are strongest during the afternoon when convective activity has reached its greatest vertical extent and mixes higher-speed winds from above downward into the friction layer next to the earth. The speeds often die down rapidly as sunset approaches and convection ceases, and may be quite calm through much of the night, when the surface air becomes stratified.

13 | FORMS AND DISTRIBUTIONS OF CONDENSATION AND PRECIPITATION

Condensation

The process of condensation that changes water from vapor to liquid renders the water visible. Water in gaseous form (vapor) is not visible; therefore, clouds or fog indicate that water in liquid droplet form is suspended in the air. Condensation can be brought about either by cooling the air below its dew point, as was discussed in Chapter 7, or by injecting moisture into it to cause supersaturation. In the majority of cases, condensation is brought about by cooling the air temperature.

Condensation is a fairly automatic process once the air becomes saturated with water vapor, that is, once the relative humidity reaches 100 percent. Although in theory supersaturation may result to several hundred percent relative humidity, in very calm, clean air free of condensation nuclei, in nature this really does not occur. The air contains so many hygroscopic (water-attracting) particles (such as sea salt, smoke, and dust) that condensation is usually guaranteed at about 100 percent relative humidity. In fact, in very dirty air condensation may begin when the relative humidity is no more than 90 percent.

Air can be cooled either by diabatic processes at its base or by adiabatic processes through lifting, as was discussed in Chapter 8. Through diabatic processes heat is lost by the air to its external environment,

such as the surface of the earth and objects on it. If the air's temperature is cooled below the dew point, some visible form of condensation will occur. Such cooling is usually accomplished by either of two methods: (a) *radiational* heat loss from the earth's surface at night or (b) *advection* (horizontal movement) of air over a colder surface.

Dew and Frost

Dew and frost almost always are the result of radiational heat loss from the earth's surface at night. For maximum heat loss to occur to assure the decrease of surface air temperature to below the dew point, skies should be clear and the surface air relatively calm. If clouds exist, they will probably absorb most of the terrestrial radiation and reradiate it back to earth, so the surface air temperature will not reach the dew point during the night. And if there is too much mixing of the surface air, the cooling of air in contact with the colder earth's surface will be mixed upward through too thick a layer of air for any portion of it to be cooled below its dew point. On the other hand, the radiational heat loss from the earth's surface on calm, clear nights will be great, and the cooling of the air in contact with the earth's surface will be limited to a thin layer. Under such conditions a surface air temperature inversion may be established, and dew or frost may form on surface objects.

Whether dew or frost forms depends upon the relationship between the dew point temperature and the freezing temperature. If the dew point is below freezing, the vapor in the air will change directly to tiny ice crystals without going through the liquid state. Since the ice crystals have much air trapped within them, the frost takes on a very whitish, fluffy appearance. This is known as hoar (white) frost. Of course, if the dew point is a little above freezing, the cooling during the night may reduce the air temperature, first, to below the dew point, which causes dew to form, and then further to below the freezing point, which freezes the water that has collected on plants and other objects. This is known as black frost and is generally more damaging to plants than hoar frost.

Since air cooled by radiational heat loss tends to settle into the lowest parts of the topography, dew and frost are found most frequently in low portions of the landscape. Since net radiational losses generally occur only at night, dew and frost are usually quickly evaporated after the sun comes up the next morning.

Fog

If many of the condensed water droplets remain suspended in the air, visibility will be impaired by fog. This will require enough mixing of the surface air to keep the water droplets suspended. But if mixing is too great, the surface cooling will be spread through too thick a layer for the surface air to be cooled below the dew point. Therefore, fog formation, particularly that due to cooling by radiational heat loss at night, places very stringent requirements on atmospheric motion. Because of such strictly limited criteria for fog formation, it is difficult to forecast.

Fog is categorized according to the method by which it was produced. Most

fog is due to the cooling of the surface temperature below the dew point, either through radiational heat loss or through the advection of moist air over a cooler surface. Radiation fog is usually thin and spotty, commonly occurring only in the lowest topographic depressions. It typically dissipates shortly after sunrise after insolation once again raises the temperature of the surface air above the dew point. Advection fog, on the other hand, is usually widespread over large continuous areas, thick, and persistent. Since it does not depend upon radiational heating and cooling, it does not show a diurnal regime, but persists as long as the surface air is advected over a cooler surface. This is very prevalent in some coastal areas, particularly those paralleled by cool ocean currents, such as many of the subtropical western coasts of the world. In these areas fog occurs most frequently in summer, when the temperatures of the oceans are coolest relative to those of adjacent land.

Advection may also be a factor in the formation of two other types of fog: upslope fog and warm front fog. When surface winds blow persistently up mountain slopes or up a gradual incline, as they do when they move northwestward from the Gulf Coast of the United States into the western Great Plains of eastern Colorado and Wyoming, adiabatic cooling can reduce the temperature below the dew point and clouds can form that are still on the ground, hence fog. Thus, windward slopes of mountain ranges often record high frequencies of fog.

Thick, persistent fog is often associated with leading edges of warm fronts in cyclonic storms. Warm, moist air rising over colder air produces a vapor pressure gradient directed downward from the moister air above. Since the actual vapor pressure of the warm air is higher than the saturation vapor pressure of the colder air below, moisture will move downward even though the lower air may be saturated. This supersaturates the cool surface air and some of the moisture condenses into liquid water droplets to produce fog. Also, some of the moisture in the warm upper air may be mixed downward into the cool air, and some may fall as rain droplets into the cool air below. All these processes combine to produce a very dense, persistent fog that is often mistaken as simple advection fog in such areas as the Great Lakes region of North America, when warm, moist air is being advected northward from the Gulf of Mexico into a cyclonic system. Such fog is usually strictly demarcated along the warm frontal surface, however; it is found in the cool air underneath the warm front, but not in the warm surface air equatorward from the front's surface position. If the front becomes stationary, the fog may hang in a certain belt for many hours or even days, and then suddenly clear within a few minutes as the front pushes on again. Although advection is a necessary part of its formation, in order to bring the moist air over the frontal surface, the fog itself is produced not by the surface cooling of the warm air, but by supersaturation of the cool air underneath the front by the downward injection of water vapor.

Fog may also be formed by the injection of more water vapor into almost-saturated air when cold air is advected across a relatively warmer water surface. In middle latitudes this is a frequent occurrence over lakes and rivers during early winter, before the water surfaces freeze. Because the vapor pressure of the water surface is greater than the saturation vapor pressure of the cold air above, moisture moves upward from the water surface into the air, even though the air might already be saturated. Almost immediately above the water surface, the water vapor will recondense into liquid water droplets and form fog. Since heat is also being added to the lower air, active turbulence will mix the water droplets upward into the air. Under such conditions it appears that steam is "boiling up" from the water's surface, hence the name "steam

fog." With extreme water-air temperature differences, this sort of fog may take on the appearance of low cumulus clouds based near the surface and extending a hundred meters or more into the air. Under such conditions major water bodies can be discerned from a distance by lines of menacing low clouds in an otherwise crystal-clear sky. In sparsely populated regions such as much of Canada and Siberia, most of the settlements are found in river valleys and along lake shores, so climatic records there often show unusually high fog frequencies in these areas during autumn and early winter. But these conditions do not pertain in uplands between streams.

Since fog often is fleeting in existence and scattered in distribution, it is difficult to represent accurately on a world scale, particularly over land where more of the fog is due to radiational cooling. Nevertheless, some general patterns can be discerned. Outstanding in fog occurrence are the subpolar seas along the fringes of the Arctic and Antarctica, especially during summer when open water fringes the ice caps and causes large temperature differentials. Fog is also relatively common along certain subtropical and middle-latitude coastal areas paralleled by cold ocean currents. And a great deal of northwestern Europe experiences persistent advection and frontal fogs during winter, as warm moist air from the Gulf Stream and its extension, the North Atlantic Drift, moves eastward from the Atlantic over the continental landmass.

Clouds

Although cooling at the surface can reduce the air below the dewpoint to form dew, frost, or fog, in such cases only a thin layer of surface air is cooled below the dew point and, hence, the total amount of condensation is limited. The only way a thick layer of air can be cooled enough to cause precipitation is by adiabatic lifting and cooling, as was illustrated briefly in Chapter 8. No matter how dry the surface air, if it can be lifted high enough, its temperature will eventually be reduced below the dew point and the condensation level in the atmosphere will be reached. Once this has happened, the liquid water droplets will become visible as clouds. Lifting may be accomplished either by the wind blowing up a terrain barrier, such as a mountain slope (*orographic* lifting), by advection up a frontal slope (*cyclonic* lifting), or by turbulent air currents set up by heating at the earth's surface (*convective* lifting).

Forms. The form the clouds take is dictated primarily by the stability characteristics of the rising air. If the air is stable, clouds will form if the air is forced to rise up a mountain or frontal slope above its condensation level; these clouds will be stratiform, featureless layered clouds that extend upward only as high as the air is forced to rise by some external force. If the air becomes unstable upon lifting, large vertical air currents will develop cumuliform clouds with large vertical extent and billowing shapes.

Stratus and cumulus are the two basic forms of clouds that produce precipitation. A third form, cirrus, are thin, wispy, high clouds composed mostly of ice crystals, but these do not cause precipitation. Since cirrus often mark the leading edges of cyclonic systems, they may be the harbingers of approaching storms, but long before precipitation commences the clouds will have lowered and thickened into stratus or cumulus types.

A fourth term that is applied to clouds is "nimbus." Nimbus does not refer to a particular form, but indicates only that the cloud is precipitating. The term is never used alone, but always in combination with one of the three terms indicating form. Hence, a cumulonimbus cloud is a cumulus cloud that has grown large enough to cause precipitation to occur. By this time, the cloud has usually grown upward beyond the freezing level in the atmosphere, and the top of the cloud has become glaciated and takes on an anvil form, as was described in Chapter 10. Nimbostratus clouds are featureless dark gray clouds that usually produce continuous precipitation over widespread areas, often associated with warm or occluded fronts. Weather services generally recognize about 30 cloud types, but they are all combinations of stratus, cumulus, and cirrus forms and their combinations with nimbus and height designations.

Clouds are the visible portion of the sky and lend much of the character to day-to-day weather. An understanding of their forms relative to the processes that produce them allows for a considerable understanding of the synoptic situation over a general area and perhaps some fairly accurate short-range forecasting, even without access to weather reports from surrounding areas. Clouds are very important climatically, since they greatly affect radiational exchanges between space and the earth's surface. Their tops are highly reflective to sunlight, having albedos comparable to that of fresh snow. Their water droplets are effective absorbers of heat energy, particularly the long wave radiation from the earth's surface. And they are very effective radiators of long-wave radiation. Most of the heat radiated from the earth-atmosphere system to space is radiated from cloud tops. And the amount of cloudiness seems to affect people psychologically: changeability seems to be more stimulating than either prolonged, dreary overcast or searing desert sun.

Distribution. On the average, the subpolar low pressure areas have the greatest amounts of cloud cover and the subtropical high pressure cells, the least. Clouds in these zones vary some in amount and shift latitudinally with seasons (Figures 13.1 to 13.4). They also vary somewhat from day to day according to synoptic situations, but not nearly as much as in other portions of the earth where circulation systems are less

Figure 13.1. Relative cloud cover, Northern Hemisphere, December–February. (From Global Atlas of Relative Cloud Cover, 1967–70 [U.S. Department of Commerce, Washington, 1971].)

consistent. Other gross features of cloud cover and their seasonal variations can be discerned even on these small maps. Such are the monsoonal effects on cloud cover over India and southeast Asia. Widespread cloudiness results from the southwest monsoon during summer. There is much less cloudiness during winter, although there is some intense cloud formation along the east coast of Vietnam, and to a lesser extent along southeastern India, derived from the northeast monsoon at this time of year. Another conspicuous seasonal reversal takes place over the Mediterranean Sea, which area appears quite clear during summer but is relatively cloudy during winter, although it is still less cloudy than Europe to the north or Africa to the south. In the Southern Hemisphere, conspicuous stratus cloud decks form off the coasts of Peru and northern Chile and off southwest Africa during winter (June–August), but are not

Figure 13.2. Relative cloud cover, Southern Hemisphere, December–February. (From Global Atlas of Relative Cloud Cover.*)*

as conspicuous during summer (December–February), when cumulus clouds tend to build up over the adjacent land.

It must be borne in mind that these polar projections of cloud cover are composites of data derived from satellite sensors over a period of four years, and represent only the amount of sky covered by clouds, not the cloud type or thickness. In general, the clouds in subpolar areas are stratiform that do not extend to great heights. They usually result in overcast conditions

and persist endlessly over time with very few breaks. This dreary type of weather is so prevalent in northern European Russia that the Russians have a special term for it—*pasmurno*—which means dull, dreary weather. Likewise, the cloud decks that form under subsidence inversions over the cool ocean currents off western South America, southwest Africa, and other similar coastal areas are stratiform, with thicknesses generally of no more than a few hundred meters. It is difficult to compare the climatic

Figure 13.3. Relative cloud cover, Northern Hemisphere, June–August. (From Global Atlas of Relative Cloud Cover.*)*

effects of such clouds with the great vertical development of cumulus extending upward 10,000 meters or more in places in the tropics, subtropics, and lower middle latitudes.

One must also remember that these composites of satellite-sensed light include light reflected from all surfaces below the satellites. By far the most reflective of these surfaces are cloud tops, snow, and ice. Thus, in higher middle latitudes, particularly in

the Northern Hemisphere, the fact that the subpolar areas have greater cloudiness during summer than during winter is somewhat masked by light reflected during winter from sea ice and snow. Also during winter, no sunlight falls on the polar areas so no satellite sensing of clouds was possible when these mosaics were compiled. Therefore, large whited-in circles blot out the pattern in these regions. Infrared sensing of clouds is now done routinely, day and night. Desert

Figure 13.4. Relative cloud cover, Southern Hemisphere, June–August. (From Global Atlas of Relative Cloud Cover.*)*

sand also has a fairly high reflectivity and might account for some of the seemingly high amounts of cloud cover in places such as the Sahara and Arabia.

The lower latitudes are depicted in Figures 13.5 and 13.6. The most-conspicuous pattern in the tropics is the extended band of clouds approximately 5- to 10-degrees latitude wide, extending across the Pacific and Atlantic oceans. This, of course, represents the intertropical convergence zone, and its position north of the equator throughout the year corroborates the discussion in Chapter 6 regarding the northerly position of the meteorological equator relative to the geographical equator, due to the stronger general circulation of the atmosphere in the Southern Hemisphere. Also well depicted on these composites are the extensive developments of stratus cloud decks over cold ocean areas off the subtropical western coasts of continents with

Figure 13.5. Relative cloud cover, December–February. (From Global Atlas of Relative Cloud Cover.*)*

their greater development during winter (June–August) off South America and southwest Africa, and during summer (June–August) off Baja California. Seasonal change in cloudiness over India and southeast Asia is also striking.

Precipitation

Precipitation is not as automatic a process as condensation. One can confidently expect that condensation will take place as 100 percent relative humidity is approached, but in no way can one be sure when precipitation will fall from a cloud. The initiation of this process seems to be related primarily to drop size. Most cloud droplets are quite small and, on falling, evaporate long before they reach the surface of the earth. Precipitation depends upon many of these cloud droplets coalescing into drops that become large enough to fall to earth.

Figure 13.6. Relative cloud cover, June–August. (From Global Atlas of Relative Cloud Cover.*)*

Several physical laws favor this happening. (a) The cloud droplets are in constant motion and undergo constant evaporation and recondensation within the cloud, providing ample opportunity for collisions to take place and for moisture to be transferred from one droplet to another. (b) Aiding this process is the fact that the surface vapor pressure of a drop of water is inversely related to the radius of the drop. Therefore, vapor pressure is higher around the surface of a small droplet than around the surface of a larger drop. This creates a vapor pressure gradient from small drops to large drops which enables larger drops to grow at the expense of smaller ones. (c) Greatly speeding up the process is the fact that above the freezing level in the atmosphere, supercooled water droplets and ice crystals coexist within clouds, and at the same temperature the surface vapor pressure is greater over water than over ice.

Thus, moisture tends to move from water droplets to ice crystals, and ice crystals act as nuclei around which large water drops form.

Some meteorologists contend that no precipitation can occur without ice crystals in the higher levels of the clouds; others point out that considerable amounts of precipitation fall in the tropics when clouds have not built up to the freezing level. It does appear that the availability of appropriate hygroscopic nuclei in sufficient abundance is a major criterion for precipitation to occur. This principle is the one exploited in so-called "cloud seeding" projects that attempt to trigger showers prematurely by injecting nuclei consisting of crystals of dry ice (frozen carbon dioxide) or silver iodide into clouds that contain supercooled water droplets at the right stage to coalesce around such nuclei. Nonetheless, the effectiveness of these methods is still under question.

Forms of Precipitation

Various forms of precipitation can occur, depending upon the relations between (a) the dew point and the freezing point in rising air that is causing moisture to condense, (b) the temperature of the air through which the precipitation must fall on its way to the surface, and (c) upon the temperature of the surface itself. If the dew point of the rising air is above freezing, water droplets will form as the air rises above the condensation level. If nothing happens to them as they fall to earth, they will fall as rain. They may evaporate on the way down and never reach the earth, or they may fall through a colder air mass underneath a front, whereupon they freeze into ice crystals that arrive at the earth as sleet. If the temperature of the cool lower air is around the freezing point, falling rain drops may not freeze until they reach the earth, whereupon they freeze to all surfaces to form glaze ice.

If the dew point of the rising air is below freezing, water vapor will turn directly into ice crystals above the condensation level. These grow into snowflakes in the cloud, and if nothing happens to them as they fall to earth they fall as snow. Many times they melt on the way down and fall as rain. Sometimes falling supercooled water droplets freeze to produce small, round, whitish pellets called "graupel" (sometimes called "soft hail"), which bounce when they hit the ground.

Snow does not fall everywhere on earth (see Figure 13.7). In general, it is a rare occurrence equatorward of about 30° latitude in the Northern Hemisphere and at an even higher latitude in the Southern Hemisphere, except in high mountains where it may occur at any latitude. And anywhere it does fall it soon melts, unless winter temperatures average below freezing. Many countries do not record amounts of snowfall; indeed, the Russians argue that what is important is not how much snow falls, but how much accumulates. This is not entirely true, of course, since occasional heavy, wet snows wreak havoc in areas such as the southeastern United States where, on the average during winter, there is little or no snow cover.

It is difficult to depict the distribution of snow, since snowfall data are inadequate and snow depth on the ground varies drastically over short distances. Yet a snow cover is such a significant control over various other aspects of climate that attempts have been made to map it accurately, and some climatic classification schemes have set major climatic boundaries according to whether or not there is a persistent snow cover during winter. Once a durable snow cover has been established, the entire heat exchange of the surface is altered drastically. (The effects of these altered relationships on temperature magnitudes and seasonal distributions will be taken up in the next chapter.)

An unusual form of precipitation is hail, which consists of spherical or irregular lumps of ice that generally fall during the hottest days of the year. This seemingly

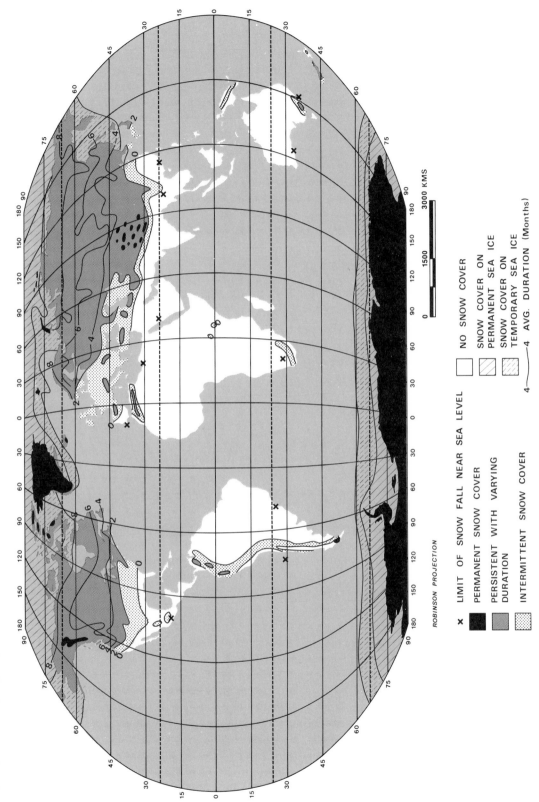

LIMIT OF SNOW FALL NEAR SEA LEVEL

× PERMANENT SNOW COVER

▨ PERSISTENT WITH VARYING
DURATION

INTERMITTENT SNOW COVER

☐ NO SNOW COVER

⧄ SNOW COVER ON
PERMANENT SEA ICE

⧄ SNOW COVER ON
TEMPORARY SEA ICE

——4—— AVG. DURATION (Months)

ROBINSON PROJECTION

0   1500   3000 KMS

Figure 13.7. Snow cover. (After G.D. Rikhter, ed., Geografiya snezhnogo pokrova [Geography of Snow Cover] [Akademii Nauk SSSR, Moscow, 1960].)

paradoxical timing is explained by the necessity for strong updrafts, which are generally present only during very warm periods when the air has a high degree of instability. Hail is formed by rain drops being carried upward beyond the freezing level. Passing through layers of air with differing cloud droplet concentration, they accumulate alternating shells of clear and rime ice. Most hailstones are less than a centimeter (.4 inches) in diameter, but hailstones as large as 14.2 centimeters (5.6 inches) in diameter have been recorded in Kansas. Hail is always associated with severe thunderstorms, and its distribution over the earth relates more to the intensity of thunderstorms than to the frequency of their occurrence (compare Figures 13.8 and 10.12). Places such as northern India with large vertical moisture gradients, hence great convective instability, are especially prone.

Amount and Distribution of Precipitation

The amount of precipitation that falls on the earth's surface during the course of a year varies greatly from one part of the earth to another, from less than 50 millimeters (2 inches) along the subtropical west coasts of South America and southwest Africa and parts of the Sahara and Central Asia, to more than 5000 mm. (200 inches) in some mountainous coastal areas, such as northeast India and Burma, the Cameroon and Guinea coasts of Africa, the western Colombian and southern Chilean coasts of South America, and spots in the Hawaiian islands (Figure 13.9).

In a few places where the atmospheric moisture content is very high and the atmospheric flow consistently converges and rises up mountain slopes, the precipitation may be copious. Such spots are Mt. Waialeale on the island of Kauai, the northwesternmost of the large islands of the Hawaiian chain, which averages 11,675 mm. (460 inches) of rainfall per year; Cherra-

punji, at an elevation of 1313 meters on the southern slopes of the Khasi Hills in northeastern India, which averages 11,419 mm. (450 inches); and Debundscha, near the base of Cameroon Mountain just north of the equator at the bend of the continent of western Africa, which averages 10,279 mm. (405 inches). Similar amounts may occur in spots in western Colombia, but comparable records are lacking there.

Areas with the least rainfall are, apparently, Arica in northern Chile, which over a period of 59 years averaged 0.75 mm. (0.03 inches), and Wadi Halfa, Sudan, which over a period of 39 years averaged 3 mm. (0.12 inches). Much of the coast of Peru and northern Chile averages less than 25 mm. (1 inch), as do portions of the southwest coast of Africa. In such places years may pass with no measurable precipitation, and then a rare shower will account for the long-term average.

Since precipitation falls in different forms, mainly rain and snow, with different proportions in different parts of the world, to construct a world distribution map (such as Figure 13.9) to show correct relative amounts of precipitation, all forms must be reduced to water equivalents. The standard practice is to convert snow to water at a ratio of 10 to 1. The wetness of snow, of course, varies considerably, and now many places weigh or melt snow samples to obtain more realistic water equivalents. In places such as the upper slopes of the southern Andes and the coastal mountains of the Alaskan Panhandle and British Columbia, much of the precipitation falls as snow. Thus, the 3000–5000 mm. of precipitation shown on the map in these areas may amount to more than 3000 centimeters (100 feet) of snow in addition to some rain. In such places glaciers accumulate in valleys between mountains, even though temperatures are not extremely cold. But so much snow falls that all cannot melt during summer, and it accumulates over the years and consolidates into glacial ice.

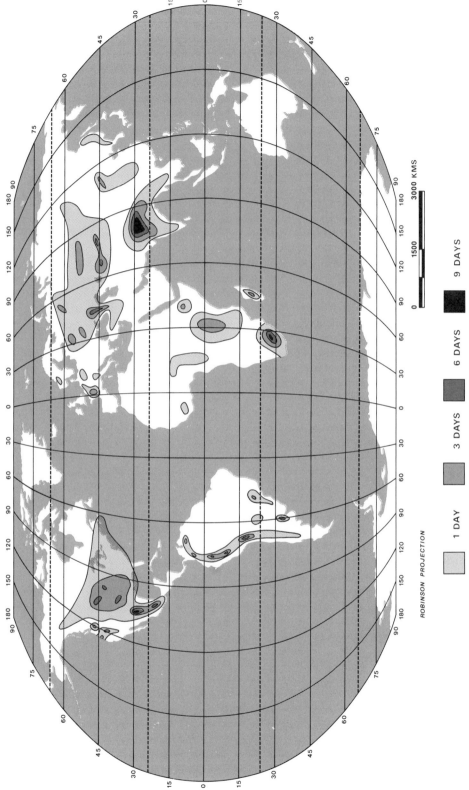

ROBINSON PROJECTION

0   1500   3000 KMS

1 DAY   3 DAYS   6 DAYS   9 DAYS

Figure 13.8. Mean annual number of days with hail. (After U.S. Army Engineer Topographic Laboratories, 1972.)

ROBINSON PROJECTION

Figure 13.9. *Mean annual precipitation, in millimeters. (From* Physical-Geographical Atlas of the World.)

It is apparent from Figure 13.9 that precipitation totals depend on two major controls: (a) amounts of moisture in the air and (b) lifting mechanisms to cause the water to precipitate out of the air. Except for a few irregularities induced by rugged topography, such as the southern Andes and northern Rockies, by far the greatest amounts of precipitation fall in near-equatorial areas where atmospheric moisture is great and the air is generally converging along portions of the intertropical convergence zone, at least part of the year. In places where orographic uplift is added, precipitation may become copious.

Areas receiving less than 500 mm. (20 inches) of precipitation can be categorized into three groups, although areally they merge in some places. First, the subtropical high pressure belts, essentially straddling the tropic circles, account for many of the extensive deserts on earth, such as the Sahara and Kalahari in Africa, the Australian desert, and the coastal desert of western South America. Particularly in the coastal areas of these deserts, the oceans provide enough moisture for precipitation to occur, but the lifting mechanism is lacking. Strong atmospheric subsidence in the subtropical high pressure cells produces adiabatic warming and, in most cases, extreme temperature inversions that stymie precipitation processes. Second, interiors of large continents, such as much of Central Asia north of the high mountains in the Soviet Union, China, and Mongolia, and the intermontane region of western United States, have meager precipitation because of the lack of a moisture source. Ordinarily, the atmosphere in these regions is relatively dry, either because they are so far from the sea or because marine air is blocked by mountain ranges. Third, the high latitudes have only meager precipitation, the polar areas generally have no more precipitation than some of the subtropical deserts. Because the air is always cold in these areas, its capacity to hold moisture is very

low, so there is little capacity for precipitation. Even though precipitation can occur fairly frequently in some of these areas, the amounts are not great. Since it is always cold or cool, the ground is either frozen or very wet during a thaw, so such areas are not classified as dry despite their meager precipitation.

Much of the plains areas of the middle latitudes receive 500–1500 mm. (20–60 inches) of precipitation, which is generally favorable for crops adapted to these areas, but seasonal and annual fluctuation in amounts may cause problems.

Seasonality. Latitudinal zones that lie within the influence of the same circulation systems throughout the year tend to have about the same precipitation in all seasons. In contrast, intermediate zones lying between two very different precipitation controls may have great seasonal contrasts, because the belts of high and low pressure, frontal zones, and so forth, shift latitudinally with seasons. Chapter 4 pointed out that the sun's direct rays shift from 23.5°N latitude on 21 June to 23.5°S latitude on 21 December, and the zone of maximum heating shifts even more widely, to about 30°–35° latitude on either side of the equator during the course of the year. Although inertias within the atmosphere keep the wind and pressure belts from shifting nearly that far, and local effects do induce irregularities, the circulation systems that largely control precipitation shift significantly from one season to another in most parts of the world.

These shifts result in the various belts shown in Figure 13.10. For instance, it can be seen that as the intertropical convergence zone shifts about 10° latitude during the year, the zone 5° or so on either side of the equator is within its influence year-round and hence receives rain throughout the year. In actuality, in much of the Atlantic and Pacific oceans, this zone remains north of the geographical equator all year, since the meteorological equator in those

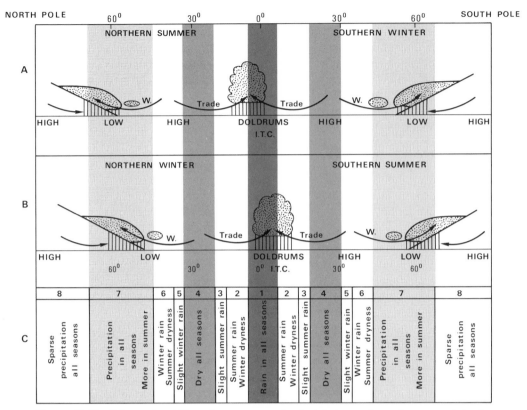

Figure 13.10. Schematic shifts of pressure and wind belts and resultant precipitation during the course of a year. (From Introduction to Meteorology *by Sverre Petterssen. Copyright © 1958 by McGraw-Hill, Inc. Used with permission of the McGraw-Hill Book Company.)*

regions lies significantly to the north of the geographical equator (see Figure 13.9). On either side of this so-called "doldrums" belt of continuous precipitation lie belts that receive rain primarily during summer, when the poleward influence of the intertropical convergence zone shifts into their vicinities. During winter the subtropical high pressure belts shift equatorward over these zones and bring drought. These are the next significant precipitation controls poleward from the intertropical convergence zone, and their influences are felt throughout the year farther poleward, around 20°–30° latitude on either side of the equator, where the climate is dry year-round.

The next significant precipitation controls poleward are the segments of the polar front with its cyclonic storms in the prevailing westerlies in the middle latitudes. These circulation systems affect belts from about 45° to about 65° latitude on either side of the equator and bring precipitation to this latitudinal zone year-round. Just to the equatorward of these belts lie zones about 10°–15° latitude wide (approximately latitudes 27°–32° to 38°–45°) that receive most of their rain during winter when the prevailing westerlies shift equatorward, but during summer they are akin to deserts as the subtropical high pressure belts shift poleward over them. Poleward of about 65° latitude there is usually some precipitation throughout the year, but in all seasons it is meager because of the restricted water-holding capacity of the colder air.

In the middle latitudes a fairly pronounced summer maximum of precipitation is the rule in interiors of continents, since at that time of year convective activity is much greater than during winter. Within these continental interiors, the summer maximum tends to lag later into summer the farther poleward the area, since in the higher middle latitudes (55°–65°) the lateness of snowmelt and ground thaw delays convective activity until later in summer. This adds a double indemnity to agriculture in higher middle latitudes: in these areas severe limitations are set by the shortness of the growing season, which is compounded by the fact that during the growing season the precipitation regime is about the opposite of what it ideally should be. In spring and early summer, the period of rapid growth of crops, when moisture is most needed it may be lacking, and later in summer and fall the harvest season is generally the wettest time of year. Since the harvest season is very short at these latitudes, this dampness may be critical, and the entire harvest may be halted by an early snowfall. Such is the case throughout much of the USSR and Canada.

In coastal areas in middle latitudes, precipitation tends to reach maximum levels in fall and early winter when sea-surface temperatures are generally at their highest, and therefore convective activity is greatest during outbreaks of relatively cold mP air masses in the rears of cyclonic storms.

Variability. In addition to seasonal cycles, precipitation amounts in many parts of the world show great degrees of aperiodic variability, with the greatest absolute differences from year to year occurring in areas of great amounts of rainfall. For instance, Cherrapunji, India, which averaged 11,421 mm. (450 inches) per year over a 74-year period, in the one year from 1 August 1860 to 31 July 1861 received 26,447 mm. (1,042 inches), more than 15,000 mm. more than normal. Obviously such large absolute deviations could not occur in regions of low rainfall that never get that much in the first place. However, on a percentage basis, generally the more arid regions of the world show the highest year-to-year variability (Figure 13.11). Thus, not only do these regions suffer from deficits of precipitation, but the little they do get cannot be relied upon. Such regions are located on the fringes of rain-bringing circulations. Some years the circulation systems penetrate fairly deeply into these dry lands and bring considerable amounts of precipitation, but in other years the circulation systems do not penetrate the regions at all and the areas remain essentially rainless.

Developers planning uses of dry lands have over the years painfully become aware of these inevitable fluctuations, and they have also come to realize that long-term averages do not represent precipitation medians. For example, if a wheat farmer can produce a crop with only 400 mm. of rainfall per year but to survive must produce a good crop every other year, he should be wary of farming in a region with a long-term yearly average of exactly 400 mm. of rain, because there will not be a 50-50 split in years above and below the 400 mm. average. There will be more dry years than wet years. Precipitation is not an infinite continuum in both directions; it cannot be less than zero. Therefore, variations on the dry side of the average cannot be as great in magnitude as variations on the wet side, and a few very wet years can balance a greater number of relatively dry years. Thus, more than half the years are going to fall below normal, and less than half above normal. If the farmer's margin of survival depends on 50 percent of the years being above average, he is going to go bankrupt.

Short-term variations of days', weeks', or months' duration occur in east-west dimensions within latitudinal belts, particularly in the middle latitudes where the standing waves of the middle troposphere so strongly influence formation and movement of surface circulation systems and

ROBINSON PROJECTION

	<10
	10–15
	16–20
	21–25
	26–40
	>40

NO DATA

0   1500   3000 KMS

Figure 13.11. Annual precipitation variability, percentage departure from normal. (After Erwin Biel.)

resultant precipitation amounts. It was pointed out in Chapter 6 that generally opposite types of weather will be experienced in opposite limbs of troughs and ridges of these wave patterns. Generally increased precipitation will be experienced on the eastward limbs of troughs (western limbs of ridges), and reduced precipitation will be experienced on western limbs of troughs (eastern limbs of ridges). Since these standing waves do migrate slowly over time, the same point on the earth's surface may find itself first under one limb for several weeks and then under another limb for another several weeks. Thus, a prolonged spell of rainy weather will be followed by a similarly long spell of dry weather, or vice versa. In some places such shifts show a considerable seasonality that brings on unusual occurrences in the weather at about the same time each year. Such phenomena are known as *singularities*. But in most cases such east-west shifts are spurious and are simply integral parts of the climate of much of the earth. Thus, even an area that would appear to be quite humid, judging from annual amounts of precipitation, could have periods of drought imbedded within the year. If such occurrences could be predicted for growing seasons several months ahead of time, it would be a great boon to agriculture, but under present technology this is impossible.

In many mid-latitude interiors of continents where so much of the precipitation depends on summer thunderstorms, great variations in precipitation amounts can occur over short distances because of the intensity of rainfall over small areas. Thus, it is difficult to generalize about drought over large areas during any given growing season.

Intensity

The manner in which precipitation falls is also very important. If it falls in infrequent cloudbursts it may do more harm than good, causing floods and heavy erosion and generally running off the surface rather than soaking in and moistening the soil. On the other hand, if it falls in very prolonged drizzly weather, it is very effective in moistening the soil and providing high humidity conditions for lush plant growth. Much of northwestern Europe receives less than 750 mm. (30 inches) of precipitation per year, but in many areas this is spread over more than 150 days, so that practically every other day gets a little rain. Under the cool, cloudy, humid conditions that prevail in this area the rainfall is quite adequate. On the other hand, in many subtropical regions precipitation falls mainly as short, heavy showers with thunderstorms, interspersed with prolonged periods of very hot, sunny weather when the earth dries up and vegetation suffers. Because of a general lack of vegetative cover in dry regions, such sharp showers run off rapidly and cause erosion and flash floods. Thus, the dry lands of the world are thrice damned: they don't get enough precipitation in the first place, it cannot be relied upon from one year to the next, and when it does come in the form of hard showers it may cause more damage than good. Agriculture in the desert areas is carried on with irrigation where water supply can be completely controlled. An unexpected heavy shower at the wrong time of year may create havoc, particularly during the harvest of perishable fruits and vegetables.

14 | TEMPERATURE

Temperature is undoubtedly the most pervasive environmental factor to which all living things respond. If one lives in an area characterized by changeable weather, probably one of the first thoughts to enter one's mind in the morning is, "what is the temperature outside?" This will determine the mode of dress and will continue to bear on activities throughout the day. Since this very important element is influenced by everything that goes on in the earth-atmosphere system, its discussion has been delayed until all the process principles have been developed.

Summary of Temperature Controls

Chapters 3–11 have illustrated that many things affect the temperature in a given place at any given time. These factors fall into the categories of either radiant heat exchange, air movement, or the change of state of water.

Heat Exchange

On a global scale the radiant heat exchange between the earth-atmosphere system and space is paramount. The magnitude of this exchange and its seasonal variations depend primarily on latitude, which, as has been shown in Chapter 4, affects both the intensity of the exchange (sun's angle above the horizon) and its duration (length of daylight period). Large-scale deviations from this latitudinal pattern of radiational exchange are induced by the state of the atmosphere and the nature of the underlying surface.

Clouds and the moisture content of the air in vapor form are the most important aspects of the state of the atmosphere that affect heat exchanges. Clouds are highly reflective to short-wave radiation from the sun and highly absorbent of long-wave radiation from the earth's surface, thereby greatly increasing the greenhouse effect. A consistently cloudy, moist atmosphere will moderate both extremes of temperature, reducing maximum temperatures during the day and preventing excessively low minimum temperatures during the night. Other constituents of the air, such as carbon dioxide and various pollutants, may have similar effects on heat exchanges, but on much smaller scales than do the various forms of moisture.

Observed temperatures are greatly affected by varying heat exchanges over different types of earth surfaces. On the broadest scale is the difference between land and water. Water surfaces change temperature much more slowly than land surfaces do, for several reasons:

1. Sunlight penetrates water to much greater depths than land, perhaps as much as 75 meters in clear water, whereas there is essentially no penetration into soil.

2. Water is free to mix, which carries the heat to even greater depths.

3. Water surfaces generally have lower

albedos than land, except perhaps at low sun angles.

4. There is usually more evaporation from an open water surface than from a land surface, which uses up more of the incident sunlight in the form of latent heat, which is then unavailable to effect a temperature change at that point on the earth's surface.

5. The specific heat of water is significantly greater than that of land materials. Specific heat is a measure of the amount of heat it takes to excite the molecules of a substance to a given energy level (temperature). Water molecules are about the most sluggish of all molecules, so a lot of heat is needed to alter the molecular motion of water. Conversely, much heat must be lost from a body of water before a significant reduction in motion takes place.

For the above reasons, the water bodies of the world, which cover three-fourths of the earth's surface, act as great heat reservoirs. They temper the extremes of temperature by absorbing great quantities of heat with minimum temperature increases during periods of positive heat balance, and by giving up great quantities of heat to the overlying atmosphere without much reduction in water temperature during periods of negative heat balance.

Similar temperature differences result, at reduced magnitudes, over varied types of land surfaces. Of course, there is little penetration of sunlight into the earth's land surface anywhere and no mixing of heat downward; but the conduction of heat downward, albedo, and evaporation rates of water vary greatly depending on whether that surface is wet soil, dry desert sand, green vegetation, snow cover, or whatever (see Figure 3.6).

During winter, large areas of the earth's surface in the middle and high latitudes are rendered essentially homogeneous to heat exchanges by their snow cover. Once a significant (30 centimeters or more) snow cover has been established, there is little connection between the surface air temperature and the soil temperature beneath the snow. The snow is a very effective insulator since it entraps much air, which benefits wintering plants by protecting them from extreme cold and from rapid fluctuations in temperature. The snow surface, on the other hand, becomes very cold because the snow reflects so much of the meager short-wave insolation incident upon it during winter, and it radiates heat away from the surface almost perfectly, as a black body, in infrared wavelengths. Since little heat is derived from the soil underneath the snow, the snow surface and the air above it become very cold. Typically, a strong surface temperature inversion develops that, in places such as Siberia and Canada, may extend upward two kilometers or more into the air. The heat exchanges between the snow surface and the inversion layer and the resulting surface temperatures during winter have already been discussed in Chapter 9.

During spring, as insolation intensity increases rapidly, surface air temperatures do not rise much above the freezing point until the snow cover has melted and the great amount of meltwater has evaporated, since the excess heat from insolation is being used in the latent heat of melting and evaporation. Thus, a snow cover delays the coming of spring. In higher latitudes a significant surface air temperature rise may be delayed almost until the summer solstice, at which time it may jump up almost instantaneously after the melting and evaporation have been accomplished, since the season of maximum heat input is then at hand.

Thus, in high latitudes, for practical purposes spring becomes nonexistent and the annual temperature curve becomes skewed, since autumn does not experience heat use for the change of state of water. In fact, as the ground and water bodies begin to freeze, autumn may be prolonged somewhat by the release of latent heat to the atmosphere through the fusion process. Thus, because of the unique heat exchange characteristics associated with a snow cover, the annual regime of temperature in high latitudes is characterized by a long, cold, coreless winter; a practically nonexistent spring; a peaked, short summer; and an autumn of reasonable length that experiences a consistently rapid decline in temperature from relatively high summer values to extremely low, early winter values.

Atmospheric Motion

Atmospheric motion, both horizontal (advection) and vertical (convection), often becomes the dominant control of temperature. Particularly in middle latitudes, air masses of widely varying characteristics in juxtaposition to one another across segments of the polar front sweep across the earth and account for nearly all the day-to-day variations in surface air temperatures. The broad-scale control by such things as latitude and land-sea differences set the background for general temperature changes that can be expected seasonally, but the daily variations superimposed upon this background are almost completely a function of the advection of air masses. These not only cause the horizontal translation of sensible temperatures from one place to another, but also greatly affect the atmospheric conditions, particularly cloudiness and humidity, which have such profound effects on insolation receipts and radiational exchanges between the earth and the atmosphere.

Cloud formations, of course, are related to vertical air motion, which also affects temperature by adiabatic heating and cooling. If the locale is a mountainous one, surface air temperature may depend more on vertical air motion than anything else. The heating effects of foehn winds have already been described in Chapters 8 and 12. They may dominate the seasonality of temperature if the general circulation over the mountainous region varies seasonally. In southern California, for instance, Los Angeles experiences its highest temperatures in September and October when a northeasterly general circulation produces foehns down the southern and western slopes of the mountains surrounding the basin on the north and east.

Altitude alone, of course, is an extremely important control on air temperature, even if the air is not in motion. As was seen in Chapters 2 and 8, in the troposphere a temperature decrease of about 6.5°C per kilometer (3.5°F per thousand feet) elevation is normal. This is usually a far greater temperature change than is experienced with any horizontal distance. Thus, if one is attempting to depict temperature distributions that are a result of latitudinal, land-sea, and other broad controls one must eliminate the altitudinal control, or it would dominate all the others and obscure the global distributions one is trying to illuminate. Therefore, many isothermal (lines of equal temperature) maps are "reduced" to sea level by adding 6.5°C per kilometer (or 3.5°F per thousand feet) to the observed temperature at the given altitude.

Change of State of Water

The latent heat involved in the changes of state of water has already been elaborated in Chapter 7 and mentioned several times in the preceding paragraphs in conjunction with other temperature controls. The influence of the presence of water in some form on observed temperatures cannot be overemphasized. Since approximately three-fourths of the earth's surface consists of standing water, and water is also present in practically all types of land surfaces and their vegetative covers, the relationships between inputs and outputs of heat and resultant temperatures are always being modified by moisture considerations. And the moisture content of the air also greatly affects the stability conditions that will, to a great extent, determine the amount of vertical motions of the air. Thus, the moisture situation of an area greatly affects observed temperatures.

Temperature Perception

The water content of the air also affects temperatures perceived by the human body, since the temperature-regulating mechanisms of the body depend to a large extent on the evaporation of perspiration from the surface of the skin. High atmospheric moisture accentuates both extremes of perceived temperature. On a hot, humid day the body is unable to lose heat as rapidly as it needs because the high humidity in the air retards the evaporation of perspiration. Therefore, one feels overheated. Conversely, on a chilly, damp day small droplets of liquid water settle out of the air onto the skin and require the body to lose more heat than it wishes to evaporate the moisture and keep the skin relatively dry. Therefore, the weather feels more chilly than the temperature really is.

Wind also causes temperatures to be perceived as chillier than they actually are because body heat is being lost more rapidly than normal. Although wind chill is difficult to measure, and no two investigators have agreed completely, a standard wind chill table has been adopted by the National Weather Service for use in the United States. This lists equivalent temperatures (with calm conditions) according to actual temperatures and wind speeds. For instance, the equivalent wind chill temperature for a recorded temperature of 0°F would be −22°F if the wind speed were 10 mph, −40°F at 20 mph, −49°F at 30 mph, −54°F at 40 mph, and −56°F at 50 mph. The relationship is a quadratic one, and wind speeds greater than 40 mph have little additional chilling effect.

Another important influence on perceived temperature is sunlight. It has been estimated that direct sunlight may make a sunbather feel as much as 14°C (25°F) warmer than would be the case with the same air temperature but cloudy conditions. Under strong sunlight, the body surface temperature rises well above the air temperature around it. This is true of any absorbing surface, whether animate or inanimate, and is a major factor in microclimatic considerations.

Because of these effects of moisture, wind, and insolation, the human body is a rather poor judge of temperature.

Expressions and Distributions of Temperature

Expressions

Temperature is observed and recorded every hour of every day of every year for thousands of climatic stations all over the world, and it is obvious that over a period of years an astronomical number of individual entries is compiled. Therefore, to comprehend and derive generalizations from the mass of data, it must be statistically reduced in some manner. Often, monthly averages are computed and isothermal maps drawn, particularly for the months having the most extreme temperatures, which in most cases are January and July. In some cases temperatures plotted on these maps are corrected to equivalent sea-level temperatures to show world distributions without the influence of altitude. To illuminate controls other than latitude, such as land-sea differences, maps showing isanomalies of temperature may be drawn to illustrate deviations of temperatures from latitudinal norms.

Mean annual ranges of temperature are computed and often plotted on maps. These show differences between mean temperatures for January and mean temperatures for July. This mean temperature range, of course, is much less than an absolute temperature range, which is the difference between the highest and lowest temperatures ever recorded in an area. On a comparative basis, however, the mean temperature range may be more realistic, since absolute maxima and minima depend, among other things, on the length of record. Mean diurnal range of temperature by month is also important. Thus, temperature data are usually represented by some sort of means and deviations. Sometimes it is important to have a finer mesh of temperature distribution with time. Such might be a frequency distribution of temperatures grouped according to size categories.

Other aspects of temperature may be important for certain purposes. For instance, the freezing point of water is a critical temperature for many things. Since many plants are killed when freezing temperatures occur, it is important to know the growing season of an area in terms of its period of consecutive frost-free days. Obviously, a crop will not grow after a killing frost, even if the temperature warms up. Also, in some cases, it might be important to know the sequence of temperature events. Wintering crops may suffer more from frequent temperature fluctuations above and below the freezing point than they would if the temperature were consistently below freezing.

Certain threshold values of temperature are important in other instances. Although most plants will not be damaged when temperatures remain above freezing, they may not grow unless the temperature rises significantly above freezing. These are threshold temperatures for growth. The degree-day concept is a way of keeping track of heat resources during the period when temperatures remain above the threshold of plant growth, which in much of the world is usually considered to be 10°C (50°F). In the United States, a 24-hour period that averages 60°F would be logged as ten growing degree-days, etc.* Similarly, heating and cooling degree-days are logs of heat resources kept for the convenience of heating and cooling houses. In the United States when outside temperature drops below 65°F, it is assumed that furnaces will turn on in houses and some fuel will be consumed. Conversely, during the warm season when the temperature outside rises above 65°F, it is assumed that air conditioning units will kick in and electricity will be consumed. Fuel and electric companies keep such logs to judge needs for energy.

Spatial Distributions

Figures 14.1 and 14.2 show world distributions of mean surface air temperatures for January and July. Probably the most conspicuous pattern is the latitudinal one. Second, land-sea differences show up as wide deviations from the latitudinal pattern, particularly in coastal areas such as over the North Atlantic northwest of Europe and in the North Pacific along the coast of North America. These land-sea contrasts are particularly distinctive during winter. For instance, the mean January temperature is about the same (−8°C) in the Barents Sea north of Scandinavia at a latitude of almost 80°N as it is in western China in the middle of the Asiatic continent at a latitude of approximately 40°N. At this time of year isotherms in northwestern Europe tend to follow coastlines more closely than parallels of latitude.

A third major influence on temperature—elevation—is also evident on these maps. Steep temperature gradients exist where the relief is great, such as in northern India where the land suddenly rises from the low, flat Ganges Plain into the Himalaya

---

*Different crops have different thresholds of growth. NOAA agencies in the United States use the 50°F threshold for corn, with an upper limit of 86°F (30°C). A day averaging about 86°F would be credited with only 36 growing degree-days.

ROBINSON PROJECTION

0   1500   3000 KMS

Figure 14.1. Mean surface air temperature in °C, January. (From Physical-Geographical Atlas of the World.)

ROBINSON PROJECTION

Figure 14.2. Mean surface air temperature in °C, July. (From Physical-Geographical Atlas of the World.)

Mountains. The intricate pattern of isotherms throughout much of southern and eastern Asia reflects the influence of topography.

A comparison of the January and July maps reveals significant latitudinal shifts in temperature zones between seasons, and highlights the fact that outside Antarctica the Southern Hemisphere, with its lack of middle-latitude landmasses, generally lacks the severity of winter that is so characteristic of much of the Eurasian and North American landmasses.

Isotherms are generally packed more closely together (temperature gradients are greater) during winter than during summer, particularly in the middle latitudes because, as was explained in Chapter 4, both the angle of sun's ray and length of day decrease poleward during winter, but during summer they decrease in opposite directions. Yet in upper middle latitudes with consistent snow covers during winter, such as much of the European plain of the Soviet Union, mean January isotherms show more north-south trend, related to distance from the sea, than they do latitudinal trend. At these latitudes during winter little insolation is received, and the small amount that is received is largely reflected from the snow's surface. Therefore, resultant temperatures are not much affected by latitude.

Figures 14.3 and 14.4 illustrate the variations of mean sea-level air temperatures from latitudinal norms during January and July. These particularly emphasize land-sea differences, which are accentuated during winter. The highest positive temperature anomaly on earth occurs at this time of year over the sea off the coast of Norway where a northeasterly extension of the Gulf Stream carries warm ocean water far poleward to warm this region. Here the January temperatures of the surface air average 26°C (47°F) above the norm for the Arctic Circle. To a lesser degree, the northeastward drift of the Japan Current in the eastern Pacific produces an abnormally warm area in the Gulf of Alaska. Extremely cold temperatures are found over the land masses in between. The far northeastern portion of the USSR experiences temperature anomalies of as much as −24°C (−43°F) during January. During July anomalies are not as great, since in the Northern Hemisphere the landmasses warm up rapidly and become a little warmer than the seas around them, while in the Southern Hemisphere during its winter season there is little land at high latitudes outside Antarctica. On the July map some of the most conspicuous anomalies are negative deviations from latitudinal norms in the summer hemisphere produced by cool ocean currents such as the California Current off the west coast of North America and the Canaries Current off northwest Africa.

Large landmasses at high latitudes show the greatest annual temperature ranges (Figure 14.5). The same conditions—distance from the sea and air stagnation in enclosed basins—that make for extreme cold in winter are conducive to abnormally high temperatures during long summer days with almost 24 hours of sunlight. By far the greatest temperature ranges on earth are found in the northeastern extremity of the Soviet Union, where the greatest negative temperature anomalies are found in January. Verkhoyansk, in a deep river valley between mountains rising to about 3000 meters on either side, at an elevation of 137 meters records a mean January surface air temperature of −48.9°C, but in July the mean temperature rises to 15.3°C. Thus, Verkhoyansk has a mean annual range of temperature of 64.2°C (115.6°F). Its absolute temperature range amounts to 103°C (185°F), from a minimum of −68°C (−90°F) to a maximum of 35°C (95°F). A little to the south in the mountains at an elevation of 740 meters, Oymyakon has experienced an absolute temperature range of 104°C, from −71 (−96°F) to +33 (91°F). As can be seen in Figure 14.6, these areas can expect minimum temperatures below

Figure 14.3. Isanomalies of mean sea-level air temperatures from latitudinal norms in °C, January. (From Physical-Geographical Atlas of the World.)

ROBINSON PROJECTION

0   1500   3000 KMS

*Figure 14.4. Isanomalies of mean sea-level air temperatures from latitudinal norms in °C, July.
(From* Physical-Geographical Atlas of the World.*)*

ROBINSON PROJECTION

Figure 14.5. Mean annual surface air temperature range (average temperature of warmest month minus average temperature of coldest month) in °C. (From Physical-Geographical Atlas of the World.)

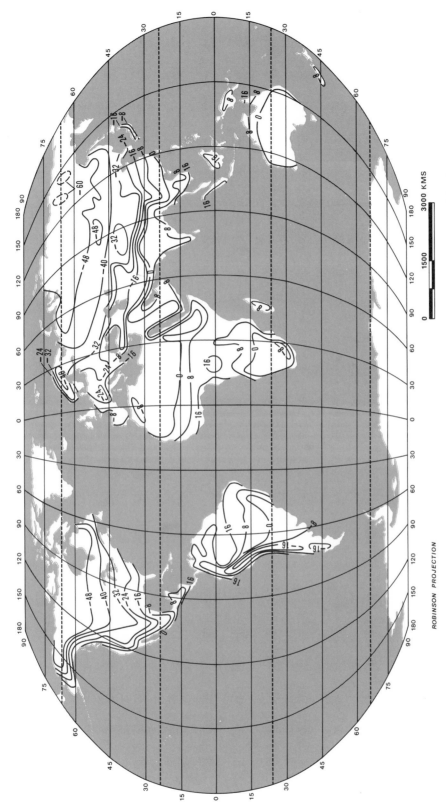

ROBINSON PROJECTION

0   1500   3000 KMS

Figure 14.6. Probable minimum surface air temperature once every ten years, in °C. (From Agroklimaticheskiy Atlas Mira [Agroclimatic Atlas of the World] [Moscow-Leningrad, 1972].)

ROBINSON PROJECTION

3000 KMS

1500

0

Figure 14.7. Days per year with maximum surface air temperatures greater than 40°C (104°F).
(From Agroclimatic Atlas of the World.)

−60°C (−76°F) at least once every ten years.

Soviet climatologists have recorded even colder temperatures in Antarctica, where at their research station, Vostok, a temperature as low as −89.2°C (−128.6°F) has been recorded. But this was at an elevation of 3488 meters above sea level. The annual range there is not as great, since summer is not warm on this high ice plateau: the warmest temperature ever recorded there was −21°C (−6°F).

The hottest temperatures on earth are found between 15° and 35° latitude, where maximum amounts of insolation are received at ground surface during the summer solstice under the influence of subtropical high pressure cells in dry desert locations (Figure 14.7). In large portions of the Sahara and Arabia, as well as in a small portion of the Indus Valley in Pakistan, more than 100 days per year experience temperatures greater than 40°C (104°F). The lower Colorado River region in the southern portion of intermontane western United States shows the greatest frequency of days with temperatures greater than 40°C at latitudes higher than 35°. The hottest temperature ever recorded is 58°C (136°F) at Azizzia, Libya, in northern Africa. A temperature nearly as high, 54°C (129°F), has been observed in Death Valley, California.

Temporal Variations

One might logically wonder why the months of extreme temperatures are January and July, rather than December and June during the solstice periods when the earth is receiving either the greatest amount or least amount of insolation from the sun. The answer, of course, lies in the fact that temperature is not the result of insolation alone, but of heat loss as well. As long as incoming radiation is greater than outgoing radiation, the temperature will rise. Although outside the tropics the rate of heat gain is the greatest during the summer

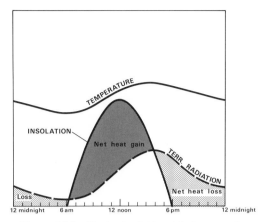

Figure 14.8. Typical relationship between insolation, terrestrial radiation, and surface air temperature during a clear day at an equinox.

solstice, there is still a net gain of heat for another month or so before terrestrial radiation begins to exceed insolation. Thus, the temperature continues to rise, albeit at a slower rate than earlier, until it reaches a peak sometime in mid-July. Conversely, during winter there is usually a net loss of heat until about mid-January.

The same sort of relationship between heat balance and temperature can be illustrated on a daily basis. Although maximum insolation occurs at noon, usually the highest temperatures during the day are observed sometime during the middle of the afternoon (Figure 14.8). Assuming sunrise at 6 AM and sunset at 6 PM, the insolation rises rapidly after sunrise to a peak at noon and falls rapidly to zero at sunset. The back radiation from the earth's surface depends upon the temperature of the earth and thus starts from a minimum value shortly after sunrise and increases, first slowly, then rapidly, and then slowly again until mid-afternoon, after which it goes into a decline. A net heat gain occurs throughout the morning and early afternoon, so the temperature continues to rise until mid-afternoon, when it reaches a maximum. After that there is a net heat loss that continues through the night, when no insolation is being received. The net loss

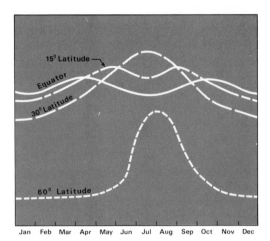

Figure 14.9. Theoretical seasonal temperature regimes and relative magnitudes at various latitudes.

continues shortly beyond sunrise, since little insolation is received from a rising sun low on the horizon. Thus, minimum temperatures generally occur shortly after sunrise.

Figure 14.8, of course, ignores all other influences on temperature such as clouds, advection, and vertical movements of the air. And it is not entirely correct for the heat exchange at the point at which the temperature is being measured, since it assumes an earth-atmosphere interface that is impervious to heat transfer downward. In actuality there is some heat transfer downward, by conduction several centimeters into soil during a day's time, and by light penetration and mixing much deeper in water. Thus, during the heat of the day some heat is stored in the soil, and later in the afternoon, as net radiation becomes negative, heat moves back upward through the soil to the earth-atmosphere interface. This slightly alters the timing of the maximum temperature. Over water it might be quite a bit, particularly on an annual basis, when heat exchange can involve a thickness of water of a hundred meters or more. Typically, along an ocean coast on an annual basis, temperature maxima may be delayed another month beyond what they are in interior land locations, so that August

is very often the month of maximum temperature in coastal areas. Over open water surfaces it may be as late as September or even October.

If controls other than heat exchanges at the surface become dominant, the temperature regime may be erratic. For instance, in southern California where foehn winds are most common in fall and early winter, the highest temperatures of the year may be experienced at that time. Near seacoasts the diurnal regime of temperature may be significantly altered by sea breezes that may begin to blow inland in late morning and continue throughout the rest of the day. This typically lops off the effects of radiational exchanges that would normally take place after noon, and leaves a reduced maximum temperature for the day that might thus occur around noon.

Within the tropics, annual regimes of temperature reflect the double passage of the direct rays of the sun during the year unless obscured by local effects. At the equator the temperature is fairly consistent throughout the entire year, since incident angle and day length never change greatly. If solar radiation is the primary control, the temperature should show two maxima—at about or shortly after the equinoxes, when the direct rays of the sun are overhead—and two minima—at about or shortly after the solstices, when the direct rays of the sun are at $23\frac{1}{2}°$ latitude on either side of the equator. Poleward the two maxima converge toward each other, since the passage of the direct rays of the sun moving poleward from the equator and returning back to the equator occur closer together in time as the tropic circles are approached. On the tropic circles the two maxima come together and remain as one single maximum and one single minimum to the poles. The amplitude between winter and summer increases poleward, and the summer period shortens as the winter season lengthens (Figure 14.9). Stations within the tropics rarely exhibit such classic re-

ROBINSON PROJECTION

■ AREAS WITHOUT FROST

0 1500 3000 KMS

Figure 14.10. Frost-free period, in days. (From Physical-Geographical Atlas of the World.)

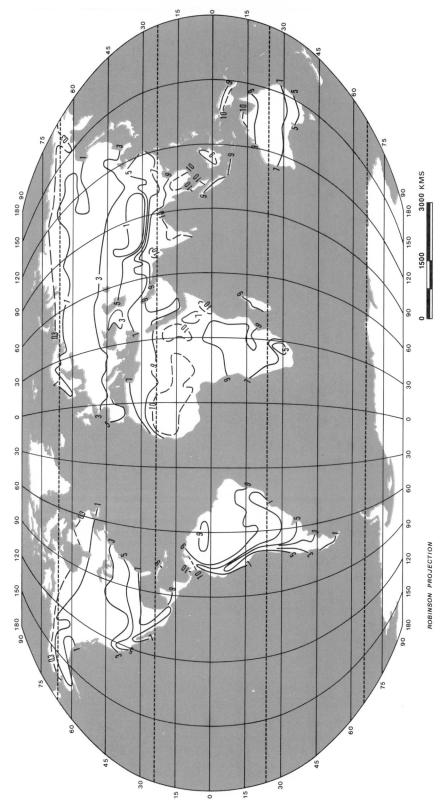

ROBINSON PROJECTION

Figure 14.11. *Average annual sums of temperatures in thousands of degrees Celsius during the period when mean daily temperatures remain above 10°C. (From Agroclimatic Atlas of the World.)*

3000 KMS
1500
0

gimes, because all sorts of things beside radiational balances, such as clouds and precipitation, may play major roles. Many tropical stations show somewhat cooler temperatures during rainy seasons and warmer temperatures during dry seasons, regardless of time of year. Since in low latitudes temperature variations are small and irregular, the designations "winter" and "summer" become meaningless. Instead the expressions "high sun" and "low sun" are often used to designate times of year when the noon sun is highest or lowest above the horizon.

Distributions of Certain Threshold Values

As was mentioned earlier, certain values of temperature act as thresholds for certain purposes. The freezing point is probably the most universally important. Figure 14.10 shows the length of frost-free period during the year. Here it can be seen that lowland areas within the tropic circles seldom, if ever, receive frosts. To about 35° latitude on either side of the equator, frosts occur only on occasional years. Beyond about 35° latitude, frosts occur every year, and of course they become more severe poleward. Along the northern fringes of North America and Eurasia, most places have less than a 60-day period when frosts ordinarily do not occur, and some years frosts may occur

in mid-summer. Over much of the Arctic and the Greenland and Antarctic ice caps frost is a constant feature. In high mountain regions worldwide, frosts typically occur on most nights of the year. During summer in high mountains the freezing point may be crossed twice each day, warming above freezing during the day and cooling below freezing at night. This so-called periglacial climate of alternate freezing and thawing has profound effects on the landscape.

The length of growing season or frost-free period, of course, is important to agriculture. But also important is the warmth of the frost-free period. To derive some measure for that, the Soviets utilize an index that is derived by adding all temperatures in degrees Celsius during the period when the mean daily temperature remains above 10°C, which they consider to be the threshold value for plant growth. These sums of temperatures are represented in Figure 14.11 and afford comparisons regarding heat resources in different parts of the world. Sums range from less than 300 along the Arctic coasts of Eurasia and North America (and even zero for the ice caps) to more than 10,000 in portions of the tropics. Minimum values of sums of temperatures have been determined for different types of crop growth and are used in the Soviet Union similar to the way the degree-day is used for agriculture in the United States and some other countries.

15 | CLIMATIC DISTRIBUTION

Classification Schemes

Most fields of science, and some other disciplines as well, find it necessary to categorize phenomena into classification systems that convert incomprehensible masses of individual pieces of information into smaller groups or relationships. Such sciences as physics and chemistry accomplish this by establishing formulas that are based on laws or principles, each of which hold true for a given body of actions or events. Other sciences, such as the biological and earth sciences, have developed hierarchical systems of categories to group, in some logical fashion, all the phenomena that fall in their purviews of study.

Some of these categorizations have real meaning, such as the biological classification of plants and animals based upon physiological functions and evolutionary delineations, and chemistry's periodic table of the elements, based upon atomic weights and numbers. Such classification schemes provide not only convenient means for conveying information, but also facilitate further research, since once the system is discovered it helps to reveal gaps in information that further research can concentrate on. Other categorizations are more arbitrary and provide no innate system to be discovered. Individual phenomena are simply grouped into categories to organize information more comprehensively. Since no inherent classification exists in these cases, a number of classifications are equally valid, each being most appropriate for its specified purpose. Such are the classifications in climatology, as well as in other fields of earth and atmospheric sciences.

The classification of climate differs from other fields of earth sciences because climate cannot be observed. It is a mental construct, composed of statistical abstractions of measured elements, on the one hand, and observable occurrences of weather episodes, on the other. To depict fully the climate of an area one must somehow convey both the end results of climate, the statistical means and deviations of measured elements, such as temperature and precipitation, and the reasons behind them—the day-to-day weather events and the atmospheric circulations, local effects, and other things that cause them.

So far discussions of areal distributions in this book have dealt with only one element at a time. Somehow these must be combined to depict the distribution of climate itself. This is not an easy task, since the climate of a place is made up of many incommensurable things, some of which are measured accurately on a regular basis, and some of which are not. This chapter will discuss some of the things that must be considered in a climatic description of an area and some of the options available for this purpose. It will culminate in an outline of a specific classification system, which will be followed by a brief description of the climate of the earth based on that system.

Empirical Versus Genetic

One must first decide whether the classification system is to be an empirical scheme, based on observations of such elements as temperature and precipitation, or genetic, based on causes behind the observed facts. Since weather episodes come and go in a few days' time, it is possible to observe, over the passage of years, many complete weather sequences that make up the climate of a region. Therefore, it might seem plausible to classify climate according to the genetic relationships of the weather episodes of which it is composed. On the other hand, a plethora of statistics have been compiled over time from observations of individual elements, such as temperature, precipitation, wind, and so forth. Therefore, good, clean data are available on which to base a rigorous and empirical classification scheme. Both types have been attempted, but generally the genetic classifications have been less successful, since they incorporate features that cannot be precisely measured and therefore must be handled subjectively. It is generally agreed that it is better to base a classification scheme on observed fact, rather than on assumed genetic relationships, and then to discuss the causes behind the facts, rather than incorporating them into the classification scheme itself.

Criteria

Next must be decided the criteria upon which the classification will be based. Climate offers a wide range of possibilities. Temperature has numerous facets, which were discussed briefly in Chapter 14. Precipitation has almost as many facets, which were discussed in Chapter 13. The various condensation forms, particularly clouds, certainly exert considerable influence on people and objects at the surface, as well as on other climatic elements. Atmospheric humidity and wind have important consequences for many things. Such elements as visibility are all-important to certain activities such as transportation. All these climatic elements are measured and recorded regularly and can be manipulated statistically.

Another group of phenomena are not as commensurable. These include variability of day-to-day weather: storminess, individual extreme happenings such as tornadoes and hurricanes, frontal passages, air mass movements, and so forth. Since no two air masses or cyclonic storms are identical, a mere counting of such things is not sufficient. Therefore they are very difficult to deal with statistically.

An adequate description of the climate of an area must use both statistics and words to include facets of both the measurable and immeasurable elements of climate. If this were not the case and all climatic information could be dealt with statistically, there would be no need for a book, since everything could be included in a cleverly devised table. A complete understanding of the climate of an area has to include both a knowledge of statistical means and deviations of measured phenomena and a knowledge of the individual weather episodes that constitute those statistics. In a study of climate, one should arm oneself with statistical compilations in one hand, and daily weather charts in the other. This is indeed how a climatic researcher studies in detail any particular climatic phenomenon or the climate of a small portion of the earth's surface.

But to provide a framework within which the climatic information for the entire earth can be presented, such as this text is trying to do, one cannot possibly handle the complexity that would result from including in a classification scheme all the measured data available, let alone the incommensurable information. Instead, a process of selection must be strictly adhered to. In most cases this has meant that classification schemes have been based only on temperature and precipitation data, while other information was covered in word descriptions about each climatic type.

The best-known and most-used classification was devised by an ethnic German from Russia, Vladimir Koeppen, during the early part of this century.[1] It is one of the simplest classifications and is based on actual degrees of temperature and millimeters of precipitation, so that final designations can easily be related back to the original data. Some investigators have criticized the Koeppen scheme for its simplicity; they say that it is not detailed enough to be of any use to researchers. This criticism, however, can be applied to any classification scheme that endeavors to provide a framework for the entire earth. No climatic researcher has found any climatic scheme to be of use in his research on small areas or limited topics, other than to provide some perspective for his own piece of work within a world context. Climatic classifications have become tools for teaching more than for research.

Some classifiers have attempted to find more meaningful indices for the thermal and moisture factors of climate than just temperature and precipitation data. The best known is the American C.W. Thornthwaite, who during the 1930s and 1940s devised a classification scheme based on what he called a "precipitation effectiveness index" and a "thermal efficiency index."[2] It is unfortunate that the derivation of these indices requires a great deal of time, and the final result is so far removed from the original data that it is difficult to relate his climatic types to actual temperatures and precipitation amounts. Also, four maps are required to depict the climate of the earth, instead of only one, as in the Koeppen system. And although the great amount of statistical manipulation creates the illusion of greater accuracy, in the final analysis Thornthwaite's classification scheme is based on no more than temperature and precipitation data. Because of its complexity, the system has not caught on in its

entirety. But certain aspects of Thornthwaite's work have thrown a great deal of light on some of the thorny problems of classification in general, and also have proved to have considerable practical value to, for example, certain types of agriculture.

The Soviet climatologists A.A. Grigoryev and M.I. Budyko, working primarily during the 1950s and 1960s, devised a scheme that uses a thermal index based upon heat units rather than temperature degrees.[3] Their thermal factor is the radiational heat balance at the earth's surface, and their moisture factor is the so-called "radiation index of aridity"—the ratio between the radiation balance at the earth's surface and the annual precipitation multiplied by the latent heat of evaporation. Although the argument for the use of heat units rather than temperature degrees is probably a valid one, in most cases radiation balance measurements are unfortunately lacking. At best, the radiation balance has to be derived as a residual in equations relating other radiational measurements, which themselves are often lacking. Therefore, Grigoryev and Budyko had to resort to the use of temperatures to depict radiation balance. They found a fairly high correlation between the annual radiation balance at the earth's surface and the sum of temperatures during the period when the mean daily temperature remains above 10°C. Therefore, their system utilizes the sum-of-temperatures concept mentioned in Chapter 14. Although this system has attracted some attention in the Soviet Union, and a climatic map of the Soviet Union has been constructed on that basis, it is not well known outside the USSR.

E.E. Federov, another Soviet climatologist working primarily during the 1920s and 1930s, was dissatisfied with the limited number of climatic elements that most classification schemes utilized, and tried to devise a system that included all facets of climate.[4] To achieve this he went back through all the climatic records for all the weather stations in the Soviet Union and characterized each day's weather with a group of symbols that included not only the measurable elements of climate, but also various incommensurables such as the synoptic situation and even the state of the surface of the ground. While his was an admirable effort and he certainly pursued this particular line of thought further than anyone else, he ended with a very complex system (that even he termed "complex climatology") that embodied some subjective decisions regarding incommensurable phenomena. Therefore, it is not completely objective, as are the three systems previously described.

In Federov's system, the climate of a place was depicted as a frequency distribution of weather types; areal distributions had to be shown one weather type at a time. Thus, the depiction of the distribution of climate over an area entailed many maps. Because of its complexity, the system as a whole has not been utilized to any extent; but like the Thornthwaite endeavor, certain aspects of the system, particularly some of the illustrative schematics, have been used in standard Soviet works such as the large *Physical Geographical Atlas of the World,* from which several maps in this text have been derived. But outside the Soviet Union the system is little known.

The four classification schemes mentioned so far are all empirical classifications. There have been attempts at genetic classification schemes, but none has been very successful. These have been based upon such things as air mass frequencies and general circulation systems, all of which are nonmeasurable and noncomparable phenomena. Usage of such criteria by necessity involves much subjective thinking, so different investigators may end up with different results even though they use the same information. Nevertheless, such efforts often have illuminated certain causal relationships that aid in explanations of climatic distributions.

Boundary Conditions

Once the decision has been made regarding the criteria to be used, boundaries need to be decided upon. If the criteria are temperature and precipitation, are any values of these elements more important than any other values? Probably not, although one might argue that the freezing point of water is a more-important temperature than any other temperature. Within the field of climate, apparently no values of such elements are more significant than any other values; therefore, to attach significance to particular values of climatic elements, classifiers have had to go outside the field of climate to some related phenomenon, such as wild vegetation or soils, to see if zones of rapid transition in these phenomena match up with any climatic values. If so, perhaps these are significant climatic boundaries, since such natural phenomena are very dependent upon climate for development. For instance, vegetation growing naturally without the interference of humans can be considered an overall climatic indicator that integrates all weather elements over the growing period, the final result of vegetative growth representing the totality of the climate of a region.

Most climatic classifications, at least the empirical ones, have utilized vegetation and other natural phenomena to determine climatic boundaries. Hence, these climatic classifications have become systems that classify the state of the surface of the earth rather than the air immediately above it. This has led to some peculiarities in climatic classifications. One is the labeling as "desert" of coastal strips such as that along the Pacific coast of Peru, which always has near-saturated surface air, much fog and low cloudiness, cool, temperate conditions with limited diurnal temperature ranges, and often a fine drizzle that keeps the streets and sidewalks damp. The only thing "desert" about such a region is its almost complete lack of measurable precipitation and its limited vegetative growth, despite its constantly cool, humid, overcast conditions. Nevertheless, on a global scale, the use of such things as natural vegetation to set climatic boundaries has alleviated much confusion and indecision and has attached some significance to categories that might otherwise have been purely arbitrary.

When Koeppen was beginning to contemplate his climatic classification, botanists were in the final stages of drawing some of the first world maps of natural vegetation, and he was impressed that the earth's vegetation was organized into fairly discrete zones separated by narrow belts of rapid transition. He reasoned that these transition belts must be caused by certain facets of climate that determined the type of plants that could grow. In his classification scheme, published in 1900, he gave names of specific types of vegetation or other related phenomena to all his climatic types. As he refined his system he narrowed his climatic boundaries to a few major ones, determined either by certain temperature values or by temperature-precipitation combinations that seemed to relate to rapid transitions of ecological groupings of plants. For instance, he found that if the temperatures for the coldest month of the year averaged below 18°C (64.4°F), many tropical plants ceased to grow. Thus, he decided that the 18°C isotherm for the coldest month was a significant boundary, the poleward limit of tropical climate. Likewise, he found that for trees to grow at high latitudes, temperatures usually had to average above 10°C (50°F) for at least one month of the year. Therefore, he decided that the 10°C average for the warmest month delineated the poleward margins of tree growth. He found that if the warmest month average was below 0°C (32°F) nothing grew, and ice caps would develop over time. Thus, he decided that the 0°C isotherm for the warmest month was the equatorward limit of ice cap climate.

Koeppen was very impressed with the effects of snow cover, since he had worked during his early life at the Main Geophysical Observatory in St. Petersburg (Leningrad), where the great Russian climatologist A.I. Voyeykov had carried out many years of fundamental research on the effects of snow cover. Koeppen decided that the equatorward edge of a persistent snow cover during winter should be a significant climatic boundary. He found that in Europe this corresponded roughly to the −3°C (26.6°F) isotherm for the coldest-month average. (North American climatologists subsequently decided that the equatorward limit of a persistent snow cover in North America corresponded more closely to the 0°C (32°F) isotherm.)

Like other climatologists attempting to formulate classification systems, Koeppen had the greatest difficulty deciding on a humid-dry boundary. This has proved to be the major bugaboo in practically every attempt to classify climate. Certainly no one isohyet (line of equal precipitation) can serve as a boundary, since moisture availability for plant growth depends not only on precipitation inputs, but also on various moisture losses from the soil. Eventually all the precipitation that falls on the earth's surface evaporates and is transferred back into the air again. So attempts have been made to fit moisture losses into some sort of evaporation formula. After much inspection of climatic and vegetation data, Koeppen finally devised three simple equations that related evaporation losses to average annual temperatures and precipitation regimes. He recognized that not only the annual amount of precipitation was important, but also seasonal regime, since if most of the precipitation fell during the warm season more of it would be lost through evaporation and would not serve to moisten the soil. He therefore devised three formulas, one for a fairly even distribution of precipitation through the year, one for precipitation concentrated during the warm season, and one for precipitation concentrated during the cold season.

But it remained to Thornthwaite to shed the greatest light on the problem of moisture loss. He coined the word "evapotranspiration" and attempted to measure it directly by inventing so-called "evaporimeters," large vessels that he embedded in the earth and in which grew indigenous types of vegetation under controlled (measurable) water inputs and losses. He quickly came to the conclusion that it was not the actual evapotranspiration that should be balanced against precipitation, but the so-called "potential evapotranspiration"—the amount that *would* evaporate were there an unlimited amount of water to be evaporated. This quantity became his famous PE (potential evapotranspiration) index. Unfortunately, his moisture-loss considerations, like Koeppen's, take into account only temperature, and not such things as relative humidity and wind speed, which also alter evapotranspiration rates. Therefore, although his estimate of moisture losses might be one of the more accurate ones, it still leaves much to be desired. Subsequent estimates by Soviet climatologists and others have produced a world map of precipitation-evapotranspiration balance (Figure 15.1).

Another of Thornthwaite's innovations is his concept of soil moisture storage. He devised a monthly log of precipitation inputs, soil moisture changes, and evapotranspiration losses, and through this running account of monthly moisture conditions he was able to demonstrate that in many areas plants may actually experience some moisture deficits during the warm season of the year, even though it might be the rainiest season of the year. This component of his work induced Birdseye Foods to support him financially so that he could set up a climatic laboratory at Seabrook, New Jersey, to advise them when to use supplemental irrigation in the growing of critical crops, such as green peas, and other vegetables for the frozen food market. Graphical representations such as that for Raleigh, North Carolina, shown in Figure 15.2, became very useful for these purposes. These graphic representations of moisture inputs, moisture losses, and soil moisture conditions illustrated very clearly the periods of soil moisture deficits occurring right during the summer maximum-rainfall periods. Of course, Figure 15.2 represents the long-term average condition, and the moisture components vary considerably from year to year. By keeping a log of current conditions during each growing season, Thornthwaite could illustrate clearly what the soil moisture conditions were at any particular time and could recommend supplemental irrigation during periods of soil moisture deficits before these deficits had existed long enough to cause curtailment of crop yields, quality, and progression of growth stages.

The Classification Used in This Book

Many of the concepts originated by Koeppen have stood up well over time, and although the simplistic approach of his system often provides only first approximations of reality, this very simplicity and the ease of relating final results to original temperature and precipitation data have resulted in its being the most frequently adopted system for teaching climatology around the world. Like the English language, it might not be the best system on earth, but because more people know it than any other system, it has become the *lingua franca* for the field of climatology, despite its irrational symbolization. Five major types of climate are designated by the five capital letters A, B, C, D, and E. Imbedded in this group are the B climates, which he designated as all the dry climates of the world. The rest are all humid climates

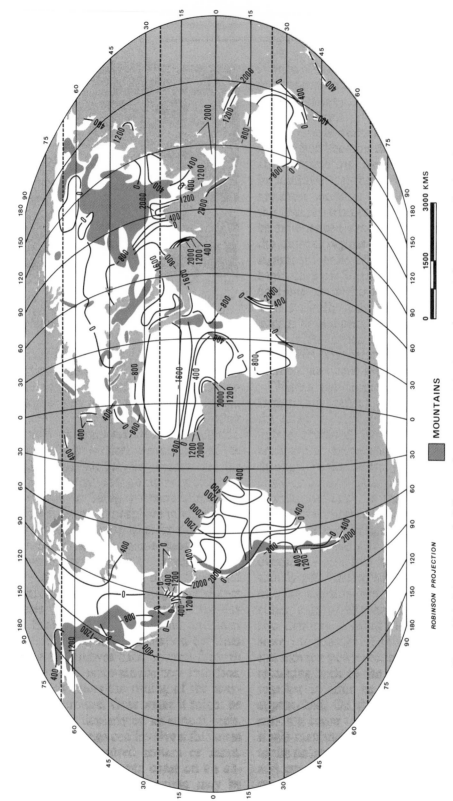

Figure 15.1. Mean annual difference between precipitation and potential evapotranspiration, in mm. (From Agroclimatic Atlas of the World.)

MOUNTAINS

ROBINSON PROJECTION

0   1500   3000 KMS

SOIL MOISTURE UTILIZATION

WATER SURPLUS

WATER DEFICIT

SOIL MOISTURE RECHARGE

Figure 15.2. Monthly moisture conditions at Raleigh, North Carolina. Bar graphs represent precipitation; the line graph, potential evapotranspiration. Potential evapotranspiration exceeds precipitation from about 10 May through about 20 September, even though summer is the season of maximum precipitation. During the last two-thirds of May and the first three-fourths of June the precipitation deficit is compensated for by plants by utilization of soil moisture (light gray area under line graph). But by 23 June all the available soil moisture is used up, and a soil moisture deficit (drought) obtains through late June, all of July, much of August, and the first two-thirds of September (dark gray area). After that a precipitation surplus is used to recharge the soil moisture (stippled area) until early December. By about 5 December the soil moisture is fully recharged, and a moisture surplus (runoff) obtains through much of December through early May (striped area). (Data from Thornthwaite, Mather, and Carter. [Modified from An Introduction to Climate, *5th ed., by Glenn T. Trewartha and Lyle H. Horn. Copyright © 1980 by McGraw-Hill, Inc. Used with permission of the McGraw-Hill Book Company].)*

and are separated from one another by certain temperature conditions. This unfortunate positioning of the symbol for dry climates in the midst of the consecutive series of letters of the alphabet, the bulk of which represent humid climates delineated on temperature bases, has had the unfortunate result of deluding many people into thinking that Koeppen's primary subdivision of climatic types is based on temperature, when actually the first-order breakdown is on the basis of moisture.[5] Only at the second level of the hierarchy does temperature come into play. It would have been more evident had the humid and dry climates been separated out more obviously; and then each subdivided on temperature bases symbolized by capital letters, or some other means.

Nevertheless, Koeppen's symbolization, probably more than any other facet of his classification, has become locked into everyone's thinking. Anyone who has studied climatology anywhere in the world knows that the A climates are tropical, the B are dry, C are mesothermal or subtropical, and so forth. Thus, this is the nomenclature that has become the *lingua franca* of climatology. Even G.T. Trewartha, who devised a classification system of his own that departs widely from the Koeppen system, has retained essentially the Koeppen nomenclature, with a few well-conceived changes and additions. Trewartha's classification, which first appeared in 1968 in the fourth edition of his *An Introduction to Climate,* will be used here, since he emphasized summer temperatures and lengths of growing seasons more than Koeppen did, and thus probably matched climatic types to such things as vegetative growth and human occupancy better than did the old Koeppen system.[6] Table 15.1 lists the climatic types, their boundary con-

Table 15.1 Climatic Classifications

Symbol	Climatic type	Boundary criteria	Rationale	Vegetation
B	**DRY**	R = 2T (s precip. regime) R = 2T + 14 (f precip. regime) R = 2T + 28 (w precip. regime) where R= hypothetical ann. precip. in cm.; T=ann. average temp. in °C; s=summer aridity (wettest winter month more than 3 × precip. of driest summer month); w=winter aridity (wettest summer month more than 5 × precip. of driest winter month); f (*feucht*)=more even distribution of precip. than in s and w.	Potential evapotranspiration exceeds precipitation (essentially the aridity limit of tree growth).	
s	Steppe	Aver. ann. precip.=R/2	Semiarid	Bunchgrass, brush, scattered trees
w	Desert (*wüste*)		Arid	Sparse shrubs, cactus
h	Hot (*heiss*)	8 mo. aver. temp.≥10°C (50°F)		
k	Cold (*kalt*)			
A	**TROPICAL HUMID**			
r	Rainy	All months aver. temp. > 18°C (65°F)		Rainforest of broadleaf evergreens
		2 mo. aver. precip.= 6 cm. (2.4 in.)		
	Wet-dry			
w	low-sun dry			Savanna grasses
s	high-sun dry			

————Coolest mo. aver. temp. = 18°C (65°F)————Poleward limit of many tropical plants

			Criteria	Vegetation
C		**SUBTROPICAL**		
	f	Humid		Mixed forests
	s	Dry summer	Wettest winter mo. 3 × driest summer mo. Driest summer mo. < 3 cm. (1.2 in.)	Chaparral, grass, live oak, eucalyptus
	w	Dry winter	Wettest summer mo. 5 × driest winter mo.	Mixed forests
	a	Hot	Warmest mo. temp. = 22°C (72°F)	
	b	Cool		
			——— 8 mo. aver. temp. ≥ 10°C (50°F) ———	
D		**TEMPERATE**		
	o	Oceanic	Persistent snow cover (one mo. or more)	Mixed, coniferous, and broadleaf forests
	c	Continental	Cold mo. aver. temp. = 0°C (32°F)	
	a	hot	Warmest mo. temp. = 22°C (72°F)	Mixed forests Primarily broadleaf deciduous
	b	cool		Mixed broadleaf and coniferous
			——— 4 mo. aver. temp. ≥ 10°C (50°F) ——— Limit of most crop cultivation	
E		**BOREAL**	——— Warmest mo. = 10°C (50°F) ——— Poleward limit of tree growth	Coniferous and small-leaved trees (taiga)
F		**POLAR**		
	t	Tundra	Warmest mo. = 0°C (32°F) Limit of vegetative growth by glacial accumulation	Mosses, lichens, sedges
	i	Ice cap		No vegetation, ice covered

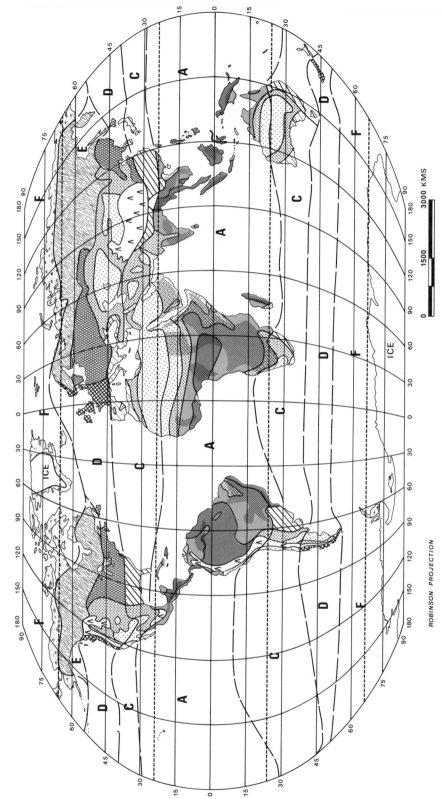

ROBINSON PROJECTION

0 1500 3000 KMS

Figure 15.3. Climatic distribution. Minor modifications have been made in southeast Australia-Tasmania and northern India. (Modified from An Introduction to Climate, 5th ed., by Glenn T. Trewartha and Lyle H. Horn. Copyright © 1980 by McGraw-Hill, Inc. Used with permission of the McGraw-Hill Book Company.)

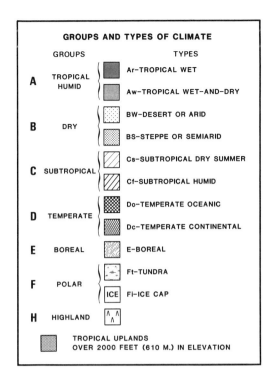

GROUPS AND TYPES OF CLIMATE

GROUPS		TYPES
A	TROPICAL HUMID	Ar–TROPICAL WET
		Aw–TROPICAL WET-AND-DRY
B	DRY	BW–DESERT OR ARID
		BS–STEPPE OR SEMIARID
C	SUBTROPICAL	Cs–SUBTROPICAL DRY SUMMER
		Cf–SUBTROPICAL HUMID
D	TEMPERATE	Do–TEMPERATE OCEANIC
		Dc–TEMPERATE CONTINENTAL
E	BOREAL	E–BOREAL
F	POLAR	Ft–TUNDRA
		Fi–ICE CAP
H	HIGHLAND	

TROPICAL UPLANDS
OVER 2000 FEET (610 M.) IN ELEVATION

ditions, and associated features. Figure 15.3 shows their distribution over the earth. Since the B climatic type is distinguished by its dryness, it will be discussed first, and then the humid types will be taken up in turn.

Climatic Types

B: Dry Climates

Climate is considered dry if the annual potential evapotranspiration exceeds the annual precipitation. From inspection of maps of natural vegetation types, temperature, and precipitation, Koeppen derived three approximate formulas, according to seasonality of precipitation, to define the humid-dry boundary, as follows:

$R = 2T$ (precipitation during wettest winter month is more than three times as much as that during the driest summer month).

$R = 2T + 14$ (precipitation evenly distributed through the year).

$R = 2T + 28$ (precipitation during the wettest summer month is more than 5 times that during the driest winter month).[7]

In these equations, R = the hypothetical annual precipitation in centimeters that would place the station with the average annual temperature T in °C exactly on the humid-dry boundary. Converted to precipitation in inches (R') and temperature in °F (T'), the formulas are as follows:

$R' = 0.44T' - 14$ (rainfall concentrated in winter)

$R' = 0.44T' - 8.5$ (rainfall evenly distributed)

$R' = 0.44T' - 3$ (rainfall concentrated in summer).[8]

Dry climates have been divided according to whether they are semiarid or arid into BS (steppe) and BW (*wüste,* the German word for desert). The boundary is taken as half that of the humid-dry boundary. The dry climates have also been divided into h (from *heiss,* the German for hot) and k (*kalt* is German for cold) according to whether more or less than eight months have average temperatures above 10°C (50°F).

To illustrate the classification of a borderline station, consider the data for Williston, North Dakota (see Table 15.2). Since the wettest summer month, June, has more than five times as much precipitation as the driest winter month, February, the formula for a w-type (winter dry) rainfall regime will be used, $R = 2T + 28$. The annual average temperature is 5°C. Therefore, $2 \times 5 + 28 = 38$ centimeters or 380 millimeters. This is the hypothetical annual precipitation that would put Williston, with an annual average temperature of 5°C, right on the humid-dry boundary. Since the annual precipitation at Williston is only 349 millimeters, the climate is on the dry side

Table 15.2   Climatic Data for Williston, North Dakota, and Los Angeles, California

	J	F	M	A	M	J	J	A	S	O	N	D	Yr.
Williston, N.D. (48°09′N, 103°37′W)													
Temperature, °C	−12	−10	−4	6	13	17	22	20	14	8	−2	−8	5
Precipitation, mm.	14	12	18	24	36	84	48	38	28	19	15	13	349
Los Angeles, Calif. (34°03′N, 118°14′W)													
Temperature, °C	13	14	15	17	18	20	23	23	22	20	17	15	18
Precipitation, mm.	78	85	57	30	4	2	0	1	6	10	27	73	373

of the humid-dry boundary. To determine whether it is steppe or desert, divide the computed R by 2 (380 ÷ 2 = 190). Since the annual precipitation at Williston is above 190, the climate is steppe rather than desert. Since fewer than eight months average above 10°C, the region is a cold steppe. Thus, the climatic classification for Williston, North Dakota, is BSk.

To illustrate the importance of the seasonality of rainfall, consider the case of Los Angeles, California, with an average annual temperature of 18°C, an annual average precipitation of 373 millimeters, and almost all of its rainfall during the winter months. In this case, the summer-dry formula, R = 2T, would be used. Thus, R = 2 × 18 = 36 centimeters or 360 millimeters. Since the annual precipitation is slightly more than that, the area is humid, even though May–September has very little rainfall. In this case the climate would classify as Csa, since the coolest month averages below 18°C, more than eight months average above 10°C, thus summers are dry, and the warmest summer month averages above 22°C.

The deserts of the world are generally located astride the tropic circles, where the subtropical high pressure cells induce subsidence much of the time, and in the interiors of large continents or behind mountain ranges, where a moisture source is lacking (see Figure 15.3). In North America and Asia dry climates from these two causes merge to produce very large areas of aridity interrupted only here and there by mountains. To some degree this is also true in the southern part of South America. More unusual locations of dry climate occur in limited areas in such places as northern Venezuela, eastern Brazil, and the eastern horn of Africa, which require other explanations that are beyond the scope of this text.[9] Altogether dry climates occupy about 26 percent of the continental areas, of which 14 percent is steppe or semiarid and 12 percent is desert or arid.

Usually, dry climatic areas are characterized by low atmospheric humidities and much sunshine, which lead to great temperature extremes on both diurnal and annual time scales. The hottest temperatures on earth have been recorded in portions of the Sahara and the below-sea-level depression of Death Valley in the American Southwest. Even at latitudes of 40° or more, daytime temperatures during summer in Soviet Central Asia can rise to 40°C or more. Williston, North Dakota, at a latitude of more than 48° has experienced a maximum temperature of 43°C. On the other hand, during winter it has experienced minima down to −46°C. Ranges between day and night at Williston average more than 14°C during summer and 10°C during winter.

A set of desert strips on coasts adjacent to cool ocean currents have very different characteristics. Such are the long coastal desert of Peru and Chile on the west coast of South America, the Namib coastal desert in southwestern Africa, the western coast of the Sahara, the coasts of southern and Baja California, and the eastern horn of Africa. In these places extreme atmospheric stability, formed both by adiabatic heating due to subsidence from above and cooling over the cool ocean currents below, produces strong temperature inversions near the surface, which restrict convective activity to the thin marine layer below the inversion. This marine layer is usually near the saturation point and is frequently filled with fog or low stratus clouds. A drizzle may even be discernible that wets the streets and causes vegetation to drip. But there is no measurable precipitation over prolonged periods of time, in some cases several years.

Arica in northern Chile averages only 0.7 millimeters rainfall per year, and this comes during a brief shower once every eight or ten years. This Atacama desert may well have the lowest measurable precipitation on earth, although large interior portions of the Sahara with no records may be just as lacking in precipitation. Nevertheless, whereas the interior portions of the Sahara are extremely hot and dry, the coastal desert of South America has consistently mild temperatures and a humid, foggy, cloudy, lower atmosphere. Arica, at approximately 18°S latitude, has mean monthly temperatures averaging between 16° and 22°C throughout the year. The hottest temperature ever recorded is 32°C, and the coldest 4°C.

Dry areas generally have sparse vegetation covers consisting of *xerophytic* (drought-resistant) species of plants whose forms reflect their physiological necessity to reduce transpiration surfaces. Middle-latitude steppes are typically vegetated by bunch grasses that may grow to heights of 30 centimeters or more, but which usually do not form a continuous sod over the surface because of the lack of sufficient soil moisture. In some places, such as large portions of the intermontane region of the American West, sagebrush and other types of woody plants may replace the grasses. Low-latitude steppes may also be vegetated by sparse grass covers, although the grasses are related more to the tropical grasses of the savannah lands on their equatorward sides, which are less nutritious than the middle-latitude grasses. Also, many tropical semiarid regions are vegetated by thorn shrubs and stunted trees.

Desert areas have even more meager vegetation covers. Usually there is much bare soil in between individual plants. In many areas, various species of brush growing a meter or two high are the predominant vegetation. In other areas, various forms of cactus are dominant. Cactus are conspicuously absent throughout most of the deserts of Central Asia, where a unique form of low tree or bush, the saxaul, has become established. In addition to the sparse perennial vegetation, a short-lived ephemeral vegetation of flowering herbaceous plants typically springs up after rainy spells. Few desert areas are completely devoid of vegetation, although this may be true in limited areas where surface materials are constantly shifted by wind, such as in some of the alluvial fan areas of the Sahara or the beach sands of the coastal deserts.

A: Tropical Humid Climates

In the tropical humid climates, all months average above 18°C (65°F). They are free of killing frosts, and therefore have year-round growing seasons. Since temperatures are always warm, the tropical climates are not subdivided on that basis, but on precipitation regime. If no more than two months have less than 6 centimeters (2.4 inches) of precipitation, the climate is considered to be rainy year-round (Ar). If more than two months have less than 6 centimeters of precipitation each, the climate is Aw if the dry season comes during winter (low-sun) or As, if it comes during summer (high-sun). In the A climates it is not realistic to talk about winter and summer, but rather low-sun, when the noon sun is lowest on the horizon, and high-sun, when it is highest above the horizon. Even low-sun and high-sun seasons may be difficult to identify, since near the equator high-sun occurs twice a year at the equinoxes, and low-sun occurs twice a year at the solstices. Also, there is often little correlation between the temperature regime and the sun regime, since the minor temperature fluctuations that exist may be caused more by the state of the sky, typically being cooler during rainy seasons than during dry seasons. Since throughout most of the tropics rain comes in the form of thundershowers that are largely the products of the intertropical convergence zone, easterly waves, and local convection, in most places maximum rainfall coincides approximately with the high-sun period. Therefore, there are only few and restricted areas of As (summer dry) climate.

The A climates are shown on Figure 15.3 in two categories, the Ar—tropical wet climate—and the Aw—tropical wet-and-dry or savanna climate. Since the earth is much larger around at the equator than farther poleward, A climates occupy a very large part of the earth's surface, approximately 36 percent. Yet this region consists more of ocean surface than land surface. A climates occupy about 43 percent of the earth's ocean surface and only 20 percent of its land surface. Therefore, A climates occupy less territory on land than dry climates do.

Ar climate. The largest areas of Ar climate straddle the equator in the Amazon basin of South America, the Congo basin of Africa, and the islands of Indonesia. The intertropical convergence zone between the trade winds generally lies in or near these areas year-round and is associated with heavy rainfall. In low-lying plains annual rainfall typically amounts to 1750–3000 or more millimeters. But in some places, such as the Colombian coast of northwestern South America and the Guinea coast of west Africa, the intertropical convergence zone is consistently located north of the equator and produces exceptionally heavy rainfalls, 5000 millimeters or more per year. The heaviest rainfalls on earth, however, are found farther poleward where consistent air flow, at least part of the year, blows against mountain coasts. Such are the northeast Indian-Bangladesh-Burmese coasts, which receive much rain from consistent southwesterly winds blowing across the Bay of Bengal during summer and early fall, and the northeast coast of some of the Hawaiian Islands which receive the northeast trades year-round. To a lesser extent the combination of rugged topography and consistency of winds also brings heavy rainfall to much of the Caribbean side of Central America and many of the Caribbean islands, eastern Madagascar Island east of South Africa, and many portions of the mountainous Philippines and surrounding island groups.

Vegetation in Ar areas is typically tropical rain forest—dense stands of huge hardwoods of many species reaching heights of 50–60 meters, which consist primarily of broadleaf evergreen trees. On the drier margins of Ar areas many trees may drop their leaves during dry spells and allow enough sunshine to get to the ground to produce junglelike tangles of undergrowth and vines.

Aw climate. The Aw climatic areas generally lie on either side of the Ar areas. The Aw is a hybrid climate, having very different precipitation controls during opposite seasons. During the high-sun season the intertropical convergence zone with its thundershowery precipitation affects the areas much as it does in the Ar areas year-round; during the low-sun season the subtropical high pressure cells shift equatorward over the areas much as they occupy the subtropical desert areas on the poleward side of Aw areas year-round.

Typically, Aw areas receive 1000–1500

millimeters of precipitation during the high-sun season but very little during the low-sun season. Thus, during the wet season the ground is saturated, the rivers are flowing full, and there is a rich vegetative growth. During the dry season the ground bakes as hard as a brick and cracks to great depths, and the vegetation withers and dies. Since most trees must have adequate moisture year-round, tree growth is sparse and coarse tropical grasses, in many cases higher than a human's head, take over. These grasslands may be interspersed with widely spaced, stunted, drought-resistant trees. In some places thorn forests of low, stunted trees take over.

As Figure 15.3 shows, large portions of Aw climatic areas lie at significant elevations above sea level. In these areas, recorded surface temperatures generally would be too cool to classify as tropical climate. Nonetheless, daily weather episodes and annual rainfall and temperature regimes reflect the tropical latitudinal positions of the areas. Therefore, the climate of these areas does not correspond to any of the other climatic groups, and on Figure 15.3 these areas are shown as having an upland variant of Aw climate.

C: Subtropical Climates

The subtropical climates have at least one month averaging below 18°C (65°F) and eight or more months averaging above 10°C (50°F). The 18°C average for the coolest month limits the growth of many tropical plants, and a monthly average of 10°C is a significant threshold of growth for many domesticated crops. Thus, the subtropical climates have at least an eight-months growing season, and most of the areas have considerably more than that. In a few places, particularly near windward coasts, frosts may never occur and the growing season is year-round. Even in some inland areas, frosts may not occur every year. On the other hand, along the poleward margins of interior areas, temperatures have dropped

as low as −25°C. In such places snow may fall frequently, occasionally in quite heavy amounts, but it does not stay on the ground long since midwinter thaws are frequent. On the other hand, windward coastal areas near the equatorward margins of subtropical climates never experience snow.

The C climates are subdivided according to two criteria: (a) the precipitation regime and (b) the summer temperature. C climates are divided into the subcategories f, s, and w, which denote seasonality of precipitation the same as they did in delimiting humid-dry boundaries. The s category of the C climates has the additional criterion that the driest month of summer must have less than 3 centimeters (1.2 inches) of precipitation. In other words, summers must be truly dry, rather than just less than one-third as wet as winter. C climates are also subdivided into a and b, according to whether or not the warmest month temperatures average above or below 22°C (72°F). Windward coasts, particularly those paralleled by cool ocean currents, generally average below 22°C during summer, while all other C climates in the interiors of continents and on leeward coasts average above 22°C. Thus, warm summers are much more widespread than cool summers.

Theoretically, then, the C climates can be subdivided into six categories—Cfa, Cfb, Csa, Csb, Cwa, and Cwb—but not all differ significantly one from another. In general, the areas having dry winters (w) also have rather cold winters, so the winter precipitation amount is of little consequence to plant growth at that time. Therefore, usually the Cf and Cw climates are grouped together. Also, the Cf and Cw climates are typically located in continental interiors or leeward coastal areas where the summers are almost always warm. Therefore the cool-summer type, if it exists at all, is usually limited to uplands. For these reasons, the four categories Cfa, Cfb, Cwa, and Cwb are grouped together in Figure 15.3 in a single pattern. Where dry winters are pronounced, such as in western China, the Cw symbol is shown, but the area is not differentiated by pattern from the Cf area farther east. Also, Csa and Csb are lumped together, since Csb generally occurs only along cool-water coasts or at elevations above sea level.

In Figure 15.3, then, the C climates are divided only into Cs and Cf climates. These two are positioned very latitudinally, generally on opposite sides of continents.

Cs climate. The Cs, or subtropical dry-summer, is usually at latitudes between 30° and 45° on western sides of continents that are dominated by the strong atmospheric subsidence of the eastern ends of subtropical highs during summer, but that during winter may occasionally be influenced by the cyclonic storms of the westerlies when the polar front swings equatorward. Hence, summers are generally dry and sunny, although in cool coastal areas there may be considerable summer fog and low stratus. Winters are still dominated to a large extent by the subtropical highs, but three or four times during winter the polar front may be positioned in the area, during which successive wave formations follow each other along the front to bring almost continuous rain for several days on end. Annual rainfall amounts typically total 375–750 millimeters in lowlands and twice that much or more in adjacent highlands, where much of the winter precipitation falls as snow.

By far the largest region on earth with Cs climate is the Mediterranean borderland, including almost all the Iberian peninsula, southern France, most of Italy, western Yugoslavia, Greece, southern Bulgaria, the coastal areas of Turkey and Palestine, foothill areas of the Middle East, and the Mediterranean fringe of Africa north of the Atlas Mountains. For this reason, the Cs climate is often called the Mediterranean climate. Nevertheless, the Mediterranean region has the least-typical Cs climate in the world. It extends farther poleward in southern Europe than anywhere else because of the influence of the warm Atlantic,

and its summers are not as dry as they are in southern California or central Chile. Much of the southern fringe of Europe does not have a simple winter maximum but, instead, complex rainfall regimes in different peninsular areas with maximum precipitation sometimes coming in fall, in fall and winter, in winter and spring, or in spring. True winter maxima occur primarily along the northern coast of Africa and the Near East as well as in the coastal areas of Portugal and southern Spain. Almost the only constant of the precipitation regime around the Mediterranean borderlands is that summer is the driest quarter of the year. But even then, appreciable precipitation may occur with thundershowers set off along the mountain slopes in the relatively warm, moist air off the Mediterranean Sea, which has much higher temperatures than the open oceans paralleling other Cs coasts.

Thus, Los Angeles, California, has a mean annual precipitation of 373 millimeters, with a maximum of 85 millimeters in February and only a trace in July. May through September average only 13 millimeters total. Athens, Greece, on the other hand, has an average annual precipitation of 402 millimeters, with a maximum of 62 millimeters in January and a minimum of 6 in June. Precipitation is distributed much more evenly through the year, with June and July the only months with very low values.

Both Los Angeles and Athens classify as Csa areas, but in Los Angeles the warmest months, July and August, average only 23°C, while in Athens they average 28°C. By contrast, Santa Monica, on a cool-water coast 25 kilometers west of downtown Los Angeles, averages only 19°C, and therefore would classify as Csb. Santa Monica experiences much more summer fog and low stratus than Los Angeles, which in turn experiences more than Athens, which is frequented more by sunny skies with occasional cumulus clouds and light showers.

In addition to the Mediterranean borderlands, southern California, and central Chile, small areas of Cs climate occur in the very southwestern tip of Africa around Capetown and in the southwestern and southeastern parts of Australia.

Since the Cs areas are frequented by long periods of aridity during summer, commonly five or six months with little or no rainfall, natural vegetation must be of the drought-resistant type. This usually means a brushy growth of many species—some evergreen, some deciduous—that collectively is known as *chaparral* in California and *maquis* in the Mediterranean region. Such vegetation has little use other than to reduce erosion and prevent mud slides that occur on the steep slopes that prevail throughout most of the Cs regions of the world. Many areas are also covered by grasses sometimes interspersed by widely spaced trees, such as the live oak in California or the cork oak in the Iberian peninsula of Europe. Eucalyptus trees are indigenous to the Cs areas of Australia and have been introduced into most of the other Cs areas of the world as ornamental trees to line streets of cities and parks, since they are able to grow to heights of 30 meters or more despite dry summers. In addition, a profusion of flowering ephemeral plants proceed rapidly through a short life cycle after the winter rainy season, much as they do after rainy seasons in many desert areas of the world.

It so happens that the Cs areas have many mountains. Therefore the climate varies locally with elevation and exposure, and heavy snow packs typically form in the mountains during winter and provide irrigation for the plains during summer. Because of the nearly year-round growing seasons and the availability of irrigation water during summer, the Cs climates have become primary areas for the growth of many crops such as citrus, for vineyards, and for a variety of other fruits and vegetables that may be exotic to other parts of the same countries. Also, the mild winters and clear, dry summers have tended to make Cs areas the playgrounds of the world for tourists.

Because of the rugged topography in most of the Cs areas, many local wind systems exist, some of such significance that they have acquired local names. The Santa Ana foehn winds of southern California are legendary and generally occur during autumn or early winter. Then the subtropical high in the eastern Pacific protrudes inland over the states of Washington and Oregon and the air circulates clockwise eastward and southward through Utah and Nevada and then south-westward across California back to the Pacific. The air, to which has been added great amounts of latent heat through the loss of precipitation on the windward slopes of mountains, funnels through passes and gaps in the San Gabriel, San Bernardino, and other mountains as it descends into the Los Angeles basin and heats adiabatically to produce surface temperatures of 35°–40°C, or even more, which are higher than anything that has been experienced during mid-summer. The highest temperature ever recorded in downtown Los Angeles is 43°C (109°F), which occurred during September. At such times it is not uncommon for even the beach cities, such as Long Beach, to experience temperatures of 38°C (100°F) or more, which contrasts starkly with maximum temperatures all summer long that hover around 20°–24°C. With such temperatures, the relative humidity often falls below 10 percent, which creates extreme fire hazards. It is typical during autumn for several fires simultaneously to rage out of control in the hills and mountains surrounding the Los Angeles basin as long as the foehn winds blow. These denude the landscape of their brushy vegetation and lay the steep slopes bare to extreme flooding, mud slides, and erosion as the winter cyclonic storms move into the area.

Similar winds descend the southern slopes of the Alps along the Riviera coast of the Mediterranean during winter, but usually the air on the Swiss plateau north of the Alps is so cold that these winds enter the coastal regions as cold winds, despite the fact that they have heated adiabatically during their descent. Their generic name, bora, comes from the region at the head of the Adriatic Sea where cold winds descend from the northeast through gaps between the eastern end of the Alps and the Dinaric Alps in western Yugoslavia. Another favorite route is along the western end of the Alps through the down-faulted Rhone valley in eastern France, where such winds are known as the "mistral."

Cf climate. As can be seen in Figure 15.3, Cf (subtropical humid) climate generally occupies southeastern portions of continents at latitudes around 25°–37°. These areas are influenced strongly by the western ends of subtropical high pressure cells which have much different characteristics than the eastern ends that influence Cs areas. In Cf areas the air circulates poleward after having crossed a broad expanse of low-latitude ocean and therefore is warm, moist, and usually conditionally unstable. Such air during summer brings prolonged sultry conditions, which is perhaps the major characteristic of this type of climate. During summer, Cf areas may be even more oppressive than equatorial areas where temperatures are not quite as high.

On the other hand, the Cf areas do have some relief when cooler, drier continental air from across the polar front occasionally invades the areas. But the continental air is more dominant during winter, and thus brings much colder winters to the Cf areas than the Cs areas experience. This is especially true in the two major Northern Hemisphere areas, southeastern United States and China. The southern Hemisphere areas, in the pampas of Argentina, Uruguay, southern Brazil, and Paraguay, the southeastern coastal area of Africa, and eastern Australia and the northern island of New Zealand, do not have major landmasses positioned poleward that can send continental polar air into these regions. The

small region of Cf climate along the eastern Black Sea in the Soviet Transcaucasus and adjacent Turkey, as well as in northern Iran bordering the southern end of the Caspian, also experience fairly mild winters, although temperatures may dip well below freezing.

Coldest occasional winter temperatures are found in southeastern United States because of the openness of the interior of the North American continent, which allows rapid passage southward of cold cP air from Canada. Near the northern margins of the Cf area there, Memphis, Tennessee, has experienced temperatures as low as −24°C (−11°F). Seven months of the year have experienced temperatures below freezing. On the other hand, nine months of the year average above 10°C, and temperatures as high as 41°C (106°F) have been experienced during summer. Combined with high relative humidity, such temperatures are truly oppressive.

Typical precipitation amounts in Cf areas range from 1200–1400 millimeters in the wetter eastern parts, to 900 millimeters or less in their drier western parts where they merge into dry climates. Typically, the maximum rainfall comes during summer in the form of thundershowers. The seasonality is very pronounced in China, where the winters tend to be quite dry because of the dominance of the northwesterly winds coming off the continent from the Asiatic High.

Interior southeastern United States is an exception to the summer maximum regime. A large area centered on the state of Tennessee has more precipitation during the winter half-year than during the summer half-year. This seems to be a function of cyclonic activity along the polar front that typically lies through this area during winter, which often brings heavy rain or wet snow when the moisture-laden air from the Gulf of Mexico overruns cold air to the north of the front. Such is not the case in eastern Asia, since during winter the polar front is usually located well out to sea, even southeast of the Japanese islands.

The Cf areas in southeastern United States and the China-Japan area experience some of the most violent weather on earth. Hurricanes and typhoons frequent the areas during summer and autumn, and in these latitudes they have usually reached their mature stages with winds often stronger than they are farther equatorward in the tropics. Also, in southeastern United States many tornadoes occur at any time of year, but especially during spring.

Since the Cf areas generally grade from quite moist conditions in their eastern portions to subhumid conditions in their western portions, the natural vegetation grades from mixed forest in the east to prairie grasses in the west. Since the pampas of South America are somewhat drier than the other Cf areas of the world, grasses were the predominant original vegetation in the area.

D: Temperate Climates

The D climatic group is characterized by four to seven months with temperatures averaging above 10°C (50°F). "Temperate" is rather a misnomer, since great extremes occur. For instance, Milwaukee, Wisconsin, located approximately in the center of the Dc region in eastern North America, has experienced a maximum temperature of 38°C (100°F) and a minimum of −31°C (−24°F). Farther away from the moderating influences of the Great Lakes and near the northwestern border of the region, Bismarck, North Dakota, has experienced temperatures ranging from 43°C (109°F) to −42°C (−44°F). Thus, the word "temperate" relates only to average conditions, not to extremes. The term is much more appropriate along western coastal areas where marine influences temper extreme conditions. And it is primarily on this basis that the D climates are subdivided into Do (temperate oceanic) and Dc (temperate continental).

The boundary between the Do and Dc climates is the 0°C (32°F) isotherm for the coldest-month average. Since the average temperature of freezing coincides closely with the equatorward limits of a snow cover in winter, the Dc areas generally have persistent snow covers for at least a month and much more in northern portions of the regions, whereas Do areas do not have persistent snow covers at lower elevations, although certainly it can snow in these regions. Also, the temperate continental areas are set off from the subtropical humid areas farther equatorward by the same snow cover conditions. This was Koeppen's original criterion to delimit the C and D climates, although he found that in Europe the equatorward edge of the persistent snow cover in winter coincided more with −3°C than with 0°C. In the classification used in this book, however, the D-C boundary has been shifted to summer growth threshold temperatures.

The D climate group can also be subdivided according to summer temperatures and precipitation regimes, the same as C climates are. But within the D climates these subdivisions are not as significant, except perhaps for the summer differentiation into Dca and Dcb, based on the boundary of 22°C for the warmest-month average temperature. A Dcs climate essentially does not exist, since in these continental locations at higher middle latitudes there is generally a pronounced summer maximum of precipitation due to added convective activity during that time of year. In fact, in some places, particularly in northern China and adjacent areas of the Far East, summers have so much more precipitation than winter that a Dcw classification would apply. But since these areas have very cold winters, the w category is insignificant and is lumped with the f category with year-round precipitation.

Dos does occur along the west coast of North America, which has unusually dry summers for such a high latitude, but the s category is not particularly significant. And most of the Do areas of the world are so marine-controlled that they do not have hot summers, so the subcategory a is not applicable to that climatic type. Thus, in Figure 15.3 only two categories, the Do and the Dc, are shown by pattern differentiation. In addition, portions of the Dc areas are divided into a and b, depending upon summer temperatures, because these exert considerable influence on types of natural vegetation and agricultural crops.

Do climate. Do climate is a very marine-controlled climate in higher middle latitudes. Its primary characteristic is its rather low annual temperature range for such a high latitude. Winters are unusually mild, and summers are relatively cool. In general precipitation is ample and evaporation rates are modest. In mountainous areas precipitation can be copious, upwards of 5000 millimeters per year, with much in the form of snow. In spite of relatively mild temperatures, the snow may not all melt during the year, and therefore mountain glaciers have accumulated in regions such as the British Columbian and Alaskan coast in western North America, the southern Andes in South America, and the western coast of the south island of New Zealand, and have produced deeply gouged, fiorded coastlines in these regions.

Northwestern Europe, the fourth and largest area of Do climate, has no very high mountains oriented perpendicular to the westerly wind flow, and therefore precipitation is generally more modest. The Do climate in western Europe stretches all the way from northern Spain through France, western Germany, the Low Countries, the British Isles, Denmark, and western Norway northward beyond the Arctic Circle. Since the mild Atlantic air has open access across the European landmass, the Do climate grades gradually into the Dc climate in central Europe, whereas in North and South America, the Do climate is limited to the western coastal areas by high mountains, across which are found dry climates on the eastern side. The south

island of New Zealand shows the same rapid decrease in precipitation from west to east, but the island is too narrow to have very dry climate on its eastern side.

All the Do areas are frequented by high relative humidities, fog, stratus clouds, and cyclonic storms. Much of the precipitation comes in the form of prolonged light-to-moderate intensity, which is very effective in moistening the soil. Therefore, even the lowlands of northwestern Europe look more humid than precipitation amounts would warrant. Consequently, all these areas originally were regions of forest, the types of which varied greatly from one continent to another. In northwestern Europe they were primarily mixed forests, while in the Pacific Northwest of North America and the southern Chilean area of South America they are primarily coniferous evergreens, some of very large size.

Dc. The Dc climate occurs only in the Northern Hemisphere, since in the Southern Hemisphere there is no land at those latitudes to produce continental polar air masses that would result in below-freezing average temperatures during winter. Three large areas cover (a) the northeastern United States and adjacent Canada, (b) central and eastern Europe, extending into southwestern Siberia, and (c) the north China-Manchuria-Korea-northern Japan area in the Far East. Although all three regions have a four-to-seven months growing season, below-freezing average temperatures during winter, and a fairly persistent snow cover during winter, there are very significant differences from place to place.

In North America the Atlantic coast portion of the region is most humid, and there precipitation amounts are similar all months of the year. Westward, the winter precipitation is less and the summer precipitation is about the same, so that the annual precipitation, which was about 1000 millimeters at New York City, is about 700 millimeters at Omaha, Nebraska, and the regime of about equal precipitation year-round in New York has changed to a pronounced summer maximum at Omaha. The greater winter precipitation in the eastern part of the region is accounted for by the convergence of winter cyclonic storm routes into the Great Lakes–St. Lawrence River region.

In the western part of the region the annual precipitation declines from around 650 millimeters in east-central Kansas northward to less than 400 millimeters in central North Dakota. This reflects the diminishing influences of the northward circulation of warm, humid air from the Gulf of Mexico around the western end of the subtropical high, which generally curves northeastward into the Great Lakes region before it reaches the northern Great Plains. But since average summer temperatures are cooler in North Dakota than in Kansas, soil moisture conditions are similar in the two regions. On occasion, however, surface temperatures have risen as high as 43°C (109°F) in Bismarck, North Dakota, almost as high as in central Kansas.

As was the case in the Cf area in southeastern United States, the Dc climatic area in North America grades from humid in the east to dry in the west. This has induced the growth of dense mixed forests in the east, westward through much of the Great Lakes region, and then mostly grasses farther west, with only an intermingling of trees, primarily in the river valleys where the groundwater table is near enough to the surface to be reached by tree roots.

The southern half of the Dc area in eastern North America has warm-month average temperatures above 22°C. Along the southern margins they average more than 26°C. This long, hot, moist growing season is conducive to the growth of a variety of crops, the main one being corn (maize). The Dca climate in North America is often referred to as the corn-belt climate. The Dcb area to the north, on the other hand, is generally too cool for corn growth

for ripe grain. Considerable corn may be grown in the southern and eastern portions of the region, but generally it is grown for silage and fodder for livestock. The region is more conducive to the growth of small grains, hay crops, and pasture. In the drier western portions of the region in North Dakota and adjacent parts of Minnesota and the prairie provinces of Canada, the primary crop is spring wheat, so that in North America the Dcb climate is often referred to as the spring wheat climate. Farther east in Wisconsin, Michigan, Ontario, and New York, however, it might well be dubbed the dairy climate, since the dairy cow has become the main source of cash for farmers whose crops (such as alfalfa, oats, and silage corn) support dairy husbandry.

As was the case in the Cf area to the south, the Dc area in North America, particularly the Dca portion, is frequented by violent thunderstorms and tornadoes during summer, which in the western half of the region occur most frequently during nighttime hours. During winter the entire Dc area usually has a considerable snow cover, which occasionally becomes quite phenomenal on leeward sides of the Great Lakes, particularly in western New York State. Except for shallow Lake Erie, the Great Lakes do not generally freeze over during winter, and cold northwesternly winds pick up much warmth and moisture as they cross the lakes, which causes enough convective activity in the lower portions of the atmosphere to produce sharp snow showers on the southeastern sides of the lakes.

Most of the Dc area in eastern Europe has cool summers, because of the dominance of air from the Atlantic. The semi-enclosed Hungarian plain, however, has hot summers, and the Dca area there has natural vegetation and crop complexes centered on corn, similar to the corn belt of North America. Because of the open access of westerly winds into Europe, the western portions of the Dc area in Europe are most moist, and the moisture supply decreases gradually eastward into western Siberia. The main moisture gradient is from north to south, northeastern Europe being humid, and southeastern Europe being dry. This is due to the high mountains in the south and the additional landmasses which keep maritime tropical air largely away from the east European plain. Therefore, the polar front in this area is less intense than it is in eastern North America, and storms are usually less violent and precipitation considerably less. Precipitation decreases from about 600 millimeters in eastern Germany to only about 450 millimeters in the middle Volga region and even less farther east. A mixed forest dominates the triangular area of the Dc climate in central and eastern Europe, which grades into grasslands toward the southeast and eventually desert in central Asia. There is a fairly pronounced summer maximum of precipitation throughout the region that grades from early summer (June) in the south to late summer (August) in the north.

The summer maximum of precipitation is much more pronounced in the Dc area of the Far East where winters are very dry, due to the strong prevalence of cold northwest winds blowing out of the Asiatic High in Mongolia and southern Siberia. Beijing (Peking) receives 623 millimeters of precipitation per year, with 243 falling in July, and only 3 each in December and January. The four-month period from December through March totals only 19 millimeters (0.76 inches) of precipitation. Thus, the winter snow cover is often absent in this region. Farther north in Manchuria and adjacent parts of the Soviet Union, where winter temperatures average −20°C or below and frequently fall to −40°C or below, the lack of snow cover allows for very deep freezing of the soil and much freezing and thawing during the transition periods at either end of winter, which is very detrimental to wintering plants. Also, strong, cold winter winds add to the wind chill

of the region, even in the coastal areas. Vladivostok on the south coast of eastern USSR averages more than 8 meters per second (18 mph) wind speed all through December and January, always from the north. These cold winter winds pick up considerable heat and moisture as they cross the Sea of Japan and produce heavy snow showers on the western mountains of Hokkaido and northern Honshu islands where, during winter, snow cover commonly accumulates to depths of 100 centimeters or more.

The mountainous regions of northern Japan, Korea, the Soviet Far East, and portions of Manchuria are primarily mixed forested regions, but in the warmer summer farther south in Manchuria and northern China, soil moisture becomes rather deficient. In fact, in the lower Huang Ho (Yellow River) area, steppe climate projects clear to the east coast.

E: Boreal Climate

The boreal (northern) climate is defined as having one to three months with temperatures averaging above 10°C (50°F). Thus, it has a short, cool growing season that generally rules out agriculture but does allow the growth of a variety of coniferous trees as well as small-leaved trees, such as birch and aspen. This is the type of vegetation that the Russians refer to as the *taiga*. The taiga is one of the two remaining large virgin forests of the world, the other being the tropical rainforest, which has much different characteristics. The boreal climate has the questionable distinction of having the greatest temperature ranges on earth, from some of the coldest winters to fairly warm, though short, summers. Oymyakon in the northern portion of the Soviet Far East has experienced temperatures as high as 33°C in summer and as low as −71°C in winter.

The boreal climate is limited to the Northern Hemisphere, since no place in the Southern Hemisphere has such cold winter temperatures and yet significant summers. The climate stretches across North America from southwestern Alaska clear across central Canada to the Atlantic coast, and across northern Eurasia from Scandinavia to the Pacific. It occupies most of Siberia except for the tundra fringe to the north and the Dc and steppe fringes along the south.

Winter is by far the dominant season in the boreal climate, occupying more than half the year. Summer is short but peaked, and temperatures rise and fall rapidly during the transitional seasons, particularly during spring, which is almost nonexistent. The prolonged winter has led to the formation of permafrost over large sections of the boreal areas. Because of the constantly cool or cold air, moisture-holding capacity is low and precipitation usually meager. Oymyakon receives only 193 millimeters of precipitation per year, most of which falls from June through September. December through April receives only 31 millimeters, almost all in the form of snow. This may mean approximately 30 centimeters of dry snow, which generally accumulates on the ground all winter since there are no midwinter thaws. Winter snowfall and snow cover increase westward in Eurasia, and are much greater in eastern Canada.

F: Polar Climates

In the polar climates no month averages above 10°C (50°F), and there is essentially no tree growth. The polar climates are divided into two categories, depending on summer temperature. If the warmest month averages above freezing, a small amount of plant growth can take place, such as mosses, lichens, sedges, and dwarf willow that creep along the ground rather than stand upright. Such vegetation is known by the Finnish word *tundra,* and the climate has become known by the same name. Tundra climate

is denoted by Ft. If no months average above freezing, then glacial ice will accumulate over time, since not all the snow can melt during the brief thaws. This, then, leads to ice caps such as those on Greenland and Antarctica, and the climate is known as the ice cap climate, denoted by Fi.

The tundra climate primarily fringes the Arctic on the northern peripheries of North America and Eurasia. Small areas also fringe the ice caps on Antarctica and Greenland. Throughout the tundra and ice cap climates the precipitation is meager because of the low moisture-holding capacity of the cold air. But since the area is frozen much of the year, moisture losses are very low, and when the surface soil does thaw, much water stands on the ground because it cannot seep into the frozen subsoil.

In high latitudes sunlight itself becomes a noticeable climatic element because of its extremes of duration—none in winter and continual in summer—and its peculiarities in relation to the horizon and direction with regard to the pole. Extremely long twilight periods are characteristic.

Highland Climate

It is impossible to classify climate in highlands since there is so much local variability, depending not only on altitude but also on exposure to sun, wind, and moisture. In general, radiational effects are more pronounced in highlands where the air is thin and pure, and advection affects are reduced. Annual range in temperature decreases while diurnal range increases. Although average monthly temperatures may classify a mountain area as a certain zonal climate, daily weather types may not be typical for that zone. Therefore, there are no very realistic lowland analogs for highland climates. In highlands, air pressure becomes a significant climatic element because its rapid reduction with elevation strongly influences the physiology of humans and other living organisms.

Notes

1. For a succinct account of Koeppen's classification system, see G.T. Trewartha and L.H. Horn, *An Introduction to Climate,* 5th ed. (New York: McGraw-Hill, 1980), pp. 397–403.

2. C.W. Thornthwaite, "An Approach Toward a Rational Classification of Climate," *Geographical Review* 38 (1948):55–94.

3. A.A. Grigoryev and M.I. Budyko, "Classification of the Climates of the U.S.S.R.," *Soviet Geography: Review and Translation* 1 (1960):3–24.

4. P.E. Lydolph, "Federov's Complex Method in Climatology," *Annals of the Association of American Geographers* 49 (1959):120–44.

5. It does fit into the geographical positioning of climatic types, however. The five major categories generally progress poleward, alphabetically, from the equator.

6. Professor Trewartha has graciously allowed the use of his classification system and world climatic map, with minor modifications.

7. This modification of the original Koeppen formula seems to fit vegetation distribution better.

8. Although these three formulas have occasionally been combined into one formula utilizing half-year precipitation amounts rather than monthly amounts, thus apparently simplifying the procedure, such simplified presentations in fact have complicated the computations, since climatic tables generally list monthly rather than half-year averages.

9. For further elaboration, see Paul E. Lydolph, *The Climate of the Earth* (Totowa, N.J.: Rowman & Allanheld, 1985).

EQUIVALENT UNITS AND CONVERSIONS

Entity	International System(SI)	cgs System	English System	Conversion
Energy(heat)	Joule(J)	Calorie(cal.)		1 cal.=4.1868J
Power	Watt(W)=1 J/sec.			
Radiant Flux Density	W/m²	Cal/cm²min Langley(ly)=1 cal/cm²		1 ly/min=697.5 W/m²
Temperature	Degrees Kelvin(K)	Degrees Celsius(°C)	Degrees Fahrenheit(°F)	K=C+273 F=9/5C+32 C=5/9(F−32)
Mass	Kilogram(kg)	kg	pound (lb)	1 kg=2.2 lb.
Length		millimeter(mm) centimeter(cm)	inch(″)	1 mm=.0394″ 1 cm=.394″
	Meter(m)	meter(m) kilometer(km)	foot(′) mile(mi)	1 m=3.28′=39.37″ 1 km=.62137 mi.
Force	Neuton(N) 1N=1 kg.m/sec²			
Pressure	Kilopascal (kPa) 1 Pa=1 N/m²	millibar (mb)	inches of mercury	1 kPa=10 mb 1 inch≅34 mb
Speed	meter/second(m/s)	m/s	miles per hour(mph) knots	1 m/s=2.237 mph 1 m/s=1.943 knots
Area		km²	mile²	1 km²=.3861 mi²

FURTHER READINGS

General Books

Anthes, Richard A., Hans A. Panofsky, John J. Cahir, and Albert Rango. *The Atmosphere*. 2nd ed. Columbus, Ohio: Charles E. Merrill, 1978.

Barrett, E.C. *Climatology from Satellites*. London: Methuen, 1974.

Barry, Roger G., and Richard J. Chorley. *Atmosphere, Weather, and Climate*. 4th ed. New York: Methuen, 1982.

Barry, Roger G., and A.H. Perry. *Synoptic Climatology: Methods and Applications*. London: Methuen, 1973.

Battan, Louis J. *Fundamentals of Meteorology*. Englewood Cliffs, N.J.: Prentice-Hall, 1979.

Boucher, Keith. *Global Climate*. New York: Halsted, 1975.

Chang, Jen-hu. *Atmospheric Circulation Systems and Climates*. Honolulu: Oriental Publishing, 1972.

Cole, Franklyn W. *Introduction to Meteorology*. 2nd ed. New York: John Wiley & Sons, 1975.

Critchfield, Howard J. *General Climatology*. 4th ed. Englewood Cliffs, N.J.: Prentice-Hall, 1983.

Gedzelman, Stanley David. *The Science and Wonders of the Atmosphere*. New York: John Wiley & Sons, 1980.

Griffiths, John F., and Dennis M. Driscoll. *Survey of Climatology*. Columbus, Ohio: Charles E. Merrill, 1982.

Hidore, John Jr. *Introduction to Weather and Climate*. Champaign, Ill.: Park Press, 1985.

Landsberg, Helmut E., ed. *World Survey of Climatology*. Amsterdam: Elsevier Science. 15 volumes, 1969–present. The most comprehensive, up-to-date compendium of information on the climate of the earth.

Lockwood, John G. *World Climatic Systems*. Baltimore: Edward Arnold, 1984.

Mather, John R. *The Climatic Water Budget in Environmental Analysis*. Lexington, Mass.: Lexington Books/D.C. Heath, 1978.

Neiburger, M., J.G. Edinger, and W.D. Bonner. *Understanding Our Atmospheric Environment*. 2nd ed. San Francisco: W.H. Freeman, 1982.

Oliver, John E., and John J. Hidore. *Climatology: An Introduction*. Columbus, Ohio: Charles E. Merrill, 1984.

Petrov, M.P. *Deserts of the World*. New York: John Wiley & Sons, 1977.

Riehl, Herbert. *Introduction to the Atmosphere*. 3rd ed. New York: McGraw-Hill, 1978.

Riley, Dennis, and Lewis Spolton. *World Weather and Climate*. 2nd ed. New York: Cambridge University Press, 1981.

Roberts, Walter Orr, and Henry Landsford. *The Climate Mandate*. San Francisco: W.H. Freeman, 1979.

Rumney, George R. *Climatology and the World's Climates*. New York: Macmillan, 1968.

Saltzman, Barry, ed. *Theory of Climate*. Vol. 25, Advances in Geophysics. Orlando, Fla.: Academic Press, 1983.

Stringer, E.T. *Foundations of Climatology*. San Francisco: W.H. Freeman, 1972.

_____. *Techniques of Climatology*. San Francisco: W.H. Freeman, 1972.

Trewartha, Glenn T. *The Earth's Problem Climates*. 2nd ed. Madison: University of Wisconsin Press, 1981.

Trewartha, Glenn T., and Lyle H. Horn. *An Introduction to Climate*. 5th ed. New York: McGraw-Hill, 1980.

Circulation Systems

Dutton, J.A. *The Ceaseless Wind: An Introduction to the Theory of Atmospheric Motion*. New York: McGraw-Hill, 1976.

Gossard, E.E., and W.H. Hooke. *Waves in the Atmosphere*. Amsterdam: Elsevier, 1975

Hoskins, Brian, and Robert Pearce, eds. *Large-Scale Dynamical Processes in the Atmosphere*. Orlando, Fla.: Academic Press, 1983.

Palmen, E., and C.W. Newton. *Atmospheric Circulation Systems: Their Structure and Physical Interpretation*. New York: Academic Press, 1969.

Perry, A.H., and J.M. Walker. *The Ocean-Atmosphere System*. New York: Longman, 1977.

Tropical Climatology

Ayoade, J.O. *Introduction to Climatology for the Tropics.* New York: John Wiley & Sons, 1983.
Newell, Reginald E., John W. Kidson, Dayton G. Vincent, and George J. Boer. *The General Circulation of the Tropical Atmosphere.* Vols. 1 and 2. Cambridge: MIT Press, 1972 and 1974.
Ramage, C.S. *Monsoon Meteorology.* Orlando, Fla.: Academic Press, 1970.
Riehl, Herbert. *Climate and Weather in the Tropics.* Orlando, Fla.: Academic Press, 1979.

Meso-Scale Systems

Atkinson, B.W. *Meso-Scale Atmospheric Circulations.* Orlando, Fla.: Academic Press, 1981.
Barry, Roger G. *Mountain Weather and Climate.* New York: Methuen, 1981.
Dewey, Kenneth F. "An Analytical Study of Lake-Effect Snowfall." Ph.D. dissertation, University of Toronto, 1973.
Eagleman, Joe R. *Severe and Unusual Weather.* New York: Van Nostrand Reinhold, 1983.
Elton, Wallace M. "Forest and Lake Breeze Patterns and the LaPorte, Indiana, Rainfall Anomaly." Ph.D. dissertation, Michigan State University, 1970.

Microclimate

Geiger, Rudolf. *The Climate Near the Ground.* Cambridge: Harvard University Press, 1965.
Lee, R. *Forest Microclimatology.* New York: Columbia University Press, 1978.
Oke, T.R. *Boundary Layer Climates.* New York: John Wiley & Sons, 1978.
Rosenberg, N.J. *Microclimate: The Biological Environment.* New York: John Wiley & Sons, 1974.
Yoshino, M.A. *Climate in a Small Area.* University of Tokyo Press, 1975.

Climatic Variation

Bach, Wilfred, Jurgen Pankrath, and Jill Williams, eds. *Interactions of Energy and Climate.* Boston: Kluwer Academic Publishers, 1980.
Budyko, M.I. *Climatic Changes.* Washington, D.C.: American Geophysical Union, 1977.
———. *The Earth's Climate: Past and Future.* Orlando, Fla.: Academic Press, 1982.
Ford, Michael J. *The Changing Climate: Responses of the Natural Flora and Fauna.* London: George Allen & Unwin, 1982.
Gribbin, John, ed. *Climatic Change.* New York: Cambridge University Press, 1978.
Lamb, Hubert H. *Climate: Present, Past and Future.* Vol. 1, *Fundamentals and Climate Now.* London: Methuen, 1972.

———. *Climate: Present, Past and Future.* Vol. 2, *Climatic History and the Future.* London: Methuen, 1977.
Lockwood, John G. *Causes of Climate.* New York: Halsted, 1979.
McCormac, Billy M., and Thomas A. Seliga, eds. *Solar-Terrestrial Influences on Weather and Climate.* Hingham, Mass.: D. Reidel, 1979.
Pittock, A.B., L.A. Frakes, D. Jenssen, J.A. Peterson, and J.Q. Zellman, eds. *Climatic Change and Variability—A Southern Perspective.* New York: Cambridge University Press, 1978.
Skinner, Brian J., ed. *Climates Past and Present.* Los Altos, Calif.: William Kaufmann, 1981.
Smith, Catherine D., and Martin Parry, eds. *Consequences of Climatic Change.* University of Nottingham Dept. of Geography, 1981.

Climate and Vegetation

Bourliere, F., ed. *Tropical Savannas.* Amsterdam: Elsevier Science, 1983.
Budyko, M.I. *Global Ecology: The Circulation of Energy and of Different Types of Matter Within the Biosphere.* Moscow: Progress Publishers, 1980.
Di Castri, F., D.W. Goodall, and R.L. Specht, eds. *Mediterranean-Type Shrublands.* Amsterdam: Elsevier Science, 1981.
Ovington, J.D., ed. *Temperate Broadleaved Evergreen Forests.* Amsterdam: Elsevier Science, 1983.
Walter, Heinrich. *Vegetation of the Earth in Relation to Climate and Eco-Physiological Conditions.* New York: Springer-Verlag, 1973.
West, N.E., ed. *Temperate Deserts and Semi-Deserts.* Vol. 5 of *Ecosystems of the World.* Amsterdam: Elsevier Science, 1982.

Climate and Man

Amiran, David H.K., and Andrew W. Wilson, eds. *Coastal Deserts: Their Natural and Human Environments.* Tucson: University of Arizona Press, 1973.
Bach, W., J. Pankrath, and W. Kellogg, eds. *Man's Impact on Climate.* Amsterdam: Elsevier Science, 1979.
Breuer, G. *Weather Modification.* New York: Cambridge University Press, 1980.
Bryson, Reid A., and Thomas J. Murray. *Climates of Hunger: Mankind and the World's Changing Weather.* Madison: University of Wisconsin Press, 1977.
Glantz, Michael H., ed. *Desertification: Environmental Degradation in and around Arid Lands.* Boulder: Westview, 1977.
Gribbin, John R. *Future Weather: Carbon Dioxide, Climate, and the Greenhouse Effect.* New York: Penguin Books, 1983.
Griffiths, John F. *Climate and the Environment: The Atmospheric Impact on Man.* Boulder: Westview, 1976.

Hess, Wilmot N., ed. *Weather and Climate Modification.* New York: John Wiley & Sons, 1974.

Kellogg, William W., and Robert Schware. *Climate Change and Society: Consequences of Increasing Atmospheric Carbon Dioxide.* Boulder: Westview, 1981.

Lamb, Hubert H. *Climate, History and the Modern World.* London: Methuen, 1982.

Liss, P.S., and A.J. Crane. *Man-Made Carbon Dioxide and Climatic Change: A Review of Scientific Problems.* Norwich, England: Geo Books, 1983.

Massachusetts Institute of Technology. *Inadvertent Climate Modification: Report of the Study of Man's Impact on Climate.* Cambridge: MIT Press, 1971.

Norwine, Jim, ed. *Climate and Human Ecology.* Houston: D. Armstrong, 1978.

Rotberg, Robert I., and Theodore K. Rabb. *Climate and History.* Princeton: Princeton University Press, 1981.

Seidel, Stephen, and Dale Keyes. *Can We Delay a Greenhouse Warming?* Washington, D.C.: Environmental Protection Agency, 1983.

Wigley, T.M.L., M.J. Ingram, and G. Farmer. *Climate and History: Studies in Past Climates and Their Impact on Man.* New York: Cambridge University Press, 1981.

Applied Climatology

Griffiths, John F. *Applied Climatology, An Introduction.* 2nd ed. London: Oxford University Press, 1976.

Hobbs, John E. *Applied Climatology: A Study of Atmospheric Resources.* Boulder: Westview, 1980.

Mather, John R. *Climatology: Fundamentals and Applications.* New York: McGraw-Hill, 1974.

Oliver, John E. *Climatology: Selected Applications.* New York: John Wiley & Sons, 1981.

Slater, Lloyd E., and Susan K. Levin, eds. *Climate's Impact on Food Supplies: Strategies and Technologies for Climate-Defensive Food Production.* Boulder: Westview, 1981.

Smith, Keith. *Principles of Applied Climatology.* New York: Halsted, 1975.

Taylor, James A., ed. *Climatic Resources and Economic Activity.* New York: Halsted, 1974.

Agroclimatology

Chang, Jen-hu. *Climate and Agriculture: An Ecological Survey.* Chicago: Aldine, 1968.

Cusack, David F. *Agroclimate Information for Development: Reviving the Green Revolution.* Boulder: Westview, 1982.

Jackson, I.J. *Climate, Water, and Agriculture in the Tropics.* New York: Longman, 1977.

Kovda, Victor A. *Land Aridization and Drought Control.* Boulder: Westview, 1980.

Raper, C. David, Jr., and Paul J. Kramer. *Crop Reactions to Water and Temperature Stresses, in Humid Temperate Climates.* Boulder: Westview, 1982.

Reifsnyder, W.E., and L.P. Smith, eds. *Agricultural Meteorology.* Amsterdam: Elsevier Science, 1983.

Rosenberg, Norman J., ed. *North American Droughts.* Boulder: Westview, 1978.

Urban Climate

Bryson, Reid A., and John E. Ross. "The Climate of the City." In Thomas R. Detwyler and Melvin G. Marcus, *Urbanization and Environment.* Belmont, Calif.: Duxbury, 1972.

Chandler, T.J. *Urban Climatology and Its Relevance to Urban Design.* Technical Note no. 149. Geneva: World Meteorological Organization, 1976.

Goward, Samuel N. "The Physical Nature of the Interface as a Factor in Urban Climate." Ph.D. dissertation, Indiana State University, 1978.

Landsberg, Helmut E. *The Urban Climate.* Orlando, Fla.: Academic Press, 1981.

INDEX